OUT OF THE WOODS

Out of the Woods

Seeing Nature in the Everyday

JULIA CORBETT

UNIVERSITY OF NEVADA PRESS | *Reno & Las Vegas*

University of Nevada Press | Reno, Nevada 89557 USA
www.unpress.nevada.edu
Copyright © 2018 by University of Nevada Press
All rights reserved
Cover design by TG Design
Cover art: Yudina Anna/Shutterstock; Lizard, Rvector; spider, Tanya Cart; beehive, Nina
Fedorova; spade, Best Vector Elements/Shutterstock.

Robert Michael Pyle granted permission to reprint "Letting the Flies Out" from his chapbook *Letting the Flies Out*. Gray's River, WA: The New Riverside Press, 2013.

A previous version of "Living in a Circle of Beating Hearts" is forthcoming in Miller, L. &
Smart, C. (Eds.), *Reimagining the Wild: Centennial Valley Reflections* (University of Utah Press).

An early version of "The Granddaddy of All Trash Days" was published in 2001 in *Weber
Studies* 19(1), 60-67.

Portions of "A Gardener Grows" were adapted from "Blooming" in my book *Seven Summers:
A Naturalist Homesteads in the Modern West*, copyright 2013 by The University of Utah Press.

An earlier version of "Robotic Iguanas" was published in 2003 in *Orion* 22(2), 36-38.

Library of Congress Cataloging-in-Publication Data
Names: Corbett, Julia B., author.
Title: Out of the woods : seeing nature in the everyday / Julia Corbett.
Description: Reno : University of Nevada Press, [2018] | Includes
bibliographical references. |
Identifiers: LCCN 2018008694 (print) | LCCN 2018029386 (ebook) | ISBN
9781943859887 (ebook) | ISBN 9781943859870 (pbk. : alk. paper)
Subjects: LCSH: Human ecology. | Nature--Effect of human beings on. | Urban
ecology (Biology) | Human-animal relationships.
Classification: LCC GF47 (ebook) | LCC GF47 .C66 2018 (print) | DDC
304.2--dc23
LC record available at https://lccn.loc.gov/2018008694

The paper used in this book meets the requirements of American National Standard for
Information Sciences—Permanence of Paper for Printed Library Materials, ANSI/NISO
Z39.48-1992 (R2002).

First Printing

Manufactured in the United States of America

To my sister-friend Sara

and to my students, who teach me so much.

Contents

Acknowledgments

It has certainly taken a village—and in this case, much of the current century—to produce this book. About sixteen years ago, I published two short essays with the vague feeling that these could be part of something bigger and that *everyday nature* had "legs." Only after an Environmental Humanities Research Professorship and a University of Utah sabbatical in Fall 2012 did I fully conceptualize the importance of "everyday nature" and begin writing more steadily.

After Dad read a draft of "The Big Hum," he went outside and sat on the cabin porch for a very long time, listening. When he came back in, he said, "I don't think I'll ever think about noise the same way again." A high compliment. His death in 2013 was the narrative thread for the chapter "Out of the Woods;" although he was not able to read the entire manuscript, he is so very present in these pages.

My sister and BFF Sara is the world's best friend and champion who bucks me up when I'm down, provides prospective and love, and makes me laugh. She read several chapters early on and believed in my approach.

My great and loyal girlfriends are *in* this book very literally; our conversations and experiences have been grist for anecdotes in several chapters, for which I thank them all. Debora Threedy graciously read and gave spot-on feedback for a great many of the chapters. It's a gift to have someone who can give both positive comments and honest critique.

Dear friend Flo Shepard is a force of nature, generous with words and actions, so enthusiastic about my writing, and though she might forget a recent remark, she never forgets the name of a bird. I can only hope to have a fraction of the impact she had on her students or to publish a book at age ninety. And I wish I could have met her late husband, Paul, whose writing I so admire.

My colleagues in the University of Utah's Environmental Humanities Graduate Program—particularly Brett Clark and Jeff McCarthy—have provided a great deal of encouragement and positive feedback for my writing. I am so happy the EH faculty are my "peeps" and that Environmental Humanities is now half of my academic home.

Special thanks to Don Feener, biology professor at the University of Utah, who graciously participated in "Brushing Them Aside" by helping me trap and identify the invertebrates in my backyard. Also special thanks to Robert Michael Pyle, who donated his time for an interview and granted permission to close the chapter with his poem.

Fellow writers at the year-long Fishtrap workshop were great supporters of my writing, in particular Isabel Tan, who has remained a dear friend and cheerleader, and Mary Dickson. They both have amazing books that someone needs to publish!

Thanks to the Western Literature Association, which presented the Manfred Award in Creative Writing, Honorable Mention to an earlier version of "Out of the Woods," September 2016.

Thanks to Justin Race, director at the University of Nevada Press, who saw the promise in this work after reading just thirty pages. I enjoyed working with him and all the press staff.

Like Oliver Sacks said, "My religion is nature." Thus, I offer profound gratitude and thanks to the natural world around and in me. We often think solely of taking from nature; it is my hope that this book, in some small way, is my part in giving back.

OUT OF THE WOODS

☉ ☉ ☉

☉ ☉ ☉

☉ ☉ ☉

Prologue

Culture and Everyday Nature

People have been falling off cliffs, getting stuck in trees, and walking into traffic. It isn't the zombie apocalypse but a monster-catching game that people play on their cell phones. Pokémon Go is a phone-based version of the old Nintendo Pokémon that uses your phone's GPS to overlay "pocket monsters" onto real-world places. As you move around, different digital monsters appear on your phone against a cartoon rendering of where you actually are. A news columnist proclaimed that this "augmented reality" game was helping people get outside, which positively affected physical and mental health.

The summer the game was released, I watched my nephew play it; he got excited when his phone told him a Pokémon was near and emitted a satisfied "yes!" when he caught one. The game was captivating but there wasn't any imagination involved; you just waited for creatures to appear. We were in Grand Teton National Park, and his eyes were stuck on the phone screen.

What's most curious to me is how this app depicts "nature." You go outside not to enjoy what's out there, but for "wild" creatures to materialize on your phone (and they "live" only in places with cell coverage). You capture the creatures in tiny red balls, who are then tamed by Pokémon trainers, who then use them to fight (which sounds a bit dark to me). Nature is merely the backdrop for technology to augment as a stylized cartoon.

Pokémon Go is one of innumerable ways that culture colors our perceptions of what and where nature is—ways that are often peculiar and largely unquestioned. Through words, pictures, and social cues, culture imprints our beliefs about nature from a very young age. Some animals are "good" (chimpanzees and butterflies), but just as many are "bad" (snakes, pigeons, and spiders). We know exactly what culturally acceptable greenery should grow around the house. We know what dryer product will make the laundry smell as if it were dried outdoors. And if you ask most children, culture has already taught them all these things. The backdrop of culture hangs throughout our lives and affects how we think about and experience nature. Thus, it affects how we practice our everyday existence on the planet.

The dictionary—a true product of human culture—defines *nature* as the natural world that exists without or beyond humans. Nature is, the Oxford dictionary declares, "the phenomena of the physical world collectively, including plants, animals, the landscape, and other features and products of the earth, as opposed to humans or human creations."

So, huh, humans (and our stuff) are not nature. We breathe the same air as those chimpanzees, drink the same water as butterflies, eat plants from the land, and use elements from the earth to make absolutely all of our stuff, but we are somehow different and apart from nature? What a cultural set-up: we're not on the same team though we share the same planet.

That set-up includes where we think nature exists, and here, I schlep my share of cultural baggage. I live in Salt Lake City, the center of a bulging urban area that stretches north and south between two mountain ranges and supports over a million people. Though I love all the activities this city offers, I have always felt that I must escape to the nearby Wasatch Mountains to be in real nature. Somehow, the birds and greenery in my backyard seem less-than.

Herein lies my challenge: though I know that humans are part of nature, I discount the nature where I live. Is it possible to draw back the cultural curtain and see the Oz that orchestrates my view? And, can I somehow learn to consider the urban and wilder nature as differently peopled versions of the same matter?

One day in class at the University of Utah, where I'm a professor of communication and environmental humanities, I lined up an assortment of water containers and asked the students what each one communicated. The "mountain spring water" in the plastic bottle was healthier and better than tap water, they said. The glass VOSS bottle meant status; a young woman said her hairdresser gave her the bottle, and we laughed that she was paying too much for her haircut. The best "PC" choice was the metal bottle. The Styrofoam cup—well, that was just so yesterday. The compostable paper cup was cool, "but like who really does that?" Though the function of each container was to hold H_2O, they knew the specific cultural meaning attached to each one.

"So where did you learn all this?" I asked them. They could not pinpoint one source; they said it all kinda mushed together. That's cultural influence: a mushy amalgamation of cues from innumerable sources: friends and family, advertising, social media, what others do or say, entertainment, celebrities, politics, children's books, geography, and on and on. Much of what we perceive nature to be is as much—maybe more—about cultural influences and social norms as it is about our experiences with it.

This matters because it is difficult to strip away the well-worn cultural meanings from a singular encounter and your immediate reaction. Instead of thinking, *oh, here is a colorful insect in my yard*, the cultural default reminds you that OMG *there's a bug in my yard, kill it!* At the zoo, kids call animals by their Disney character's name, unable to see the creature in front of them. I feel that I know wolves—even though I have never had an up-close encounter— because they are so imbued with cultural meanings. No matter the animal and my experience with it—skunk, ladybug, whale, worm—I carry cultural expectations and stereotypes with me, and I know whether to feel afraid, happy, in awe, or grossed out.

The snowy summits of the Teton Mountains pierce a brilliant blue sky on the calendar in my kitchen; there is not a person in sight. On the milk carton, black-and-white cows graze in a verdant pasture. Fridge magnets cling to last summer's pictures of a moose, a cluster of fairy slipper orchids, and girlfriends posing at a mountain vista in Oregon. Over the fireplace hangs a pen-and-ink drawing of owl chicks. In the magazine on my coffee table, a man leaps across a house-sized boulder against an azure sky (and recommends a granola bar), and an ad for laundry detergent splashes across a lush green, flower-rich meadow. A bright green toucanet with big dark eyes perched on a mossy branch stares from the screen of my laptop.

These items tell me that nature is out there, far beyond my everyday life. Here, nature is pristine and largely peopleless and looks great—the birds and animals are fine, the air and water are pure, the landscapes lush, and the vistas unbroken.

This is not the nature I live in. The magnificent encircling mountains are often blurred by air pollution. I hear the everlasting roar of the interstate in my backyard, as well as dogs, all manner of yard equipment, sirens, and construction. At night, the house is never dark. Cement, buildings, roads—all run and rise in hard straight lines and crowd, confine, and conceal patches of earth. This is where I live and work, and I want and need nature to feel more present in my life here.

In contrast, the "wild" nature in the nearby mountains seems free of contamination by human desires and actions. Historian William Cronon calls wilderness a *moral baseline* that we use to measure ourselves against, for it tells us what the world might look like in our absence or how we can (or should) try and erase our presence. Historically and culturally we have become accustomed to seeing and valuing nature in some places and not others: "To the extent that we live in an urban-industrial civilization but at the same time pretend that our real home is in the wilderness, to just that extent we give ourselves permission to evade responsibility for the lives we actually lead. . . . By imagining that our true home is in the wilderness, we forgive ourselves the homes we actually inhabit."

The city I inhabit has much to teach me about my imagined tale of these two natures and how my fingerprints and footprints blanket them both, all the way up to the atmosphere. When cultural beliefs (from the dictionary or in my head) partition these two natures in separate boxes, it conceals the spillover: how my living-large actions in everyday nature are pick-axes on the wilder nature—the very places that help me step back from human domination and touch precious pieces of myself, pieces I want present in my city life.

〜〜〜〜

If it is possible to put cultural valuations of nature aside even for a minute, on a purely physical level, there is just one intertwined, vibrant nature. We tend to think of inanimate "things" (shoes, a book, a glass of water) as passive and inert, just sitting there being, well, things. Yet, such things are just as vital and vibrant as animate beings, argues Jane Bennett in *Vibrant Matter*. If you view *vitality* as the capacity of things to affect us humans and to act as forces with their own propensities and tendencies, a glass of water exudes vitality. The propensity of the glass is to hold liquid, and the liquid affects me, hydrates me, when I drink it. All kinds of "matter" produce effects and alter how things happen—carrots, rainstorms, rocks. The nature/culture divide we cling to is both inaccurate and unhelpful, for humans are embedded in a tangled web of acting and being acted upon. For Bennett, the fantasy that we humans are somehow in charge of all those things prevents us from seeing our connection to all matter, which all emanates from the earth.

Your cell phone with the Pokémon app has vibrant *thing power*, or the curious ability of inanimate objects to produce effects both dramatic and subtle; falling off a cliff or walking into traffic certainly qualify as dramatic effects. Even though a cell phone may seem like "dead nature," the yttrium, scandium, copper, gold, and platinum are vital to the phone's operation and are just as vibrant as when they lay underground in far-flung planetary places. And if your search for a Pokémon takes you to a landfill, you undoubtedly will see cell phones there, where they are still producing effects. About 140 million cell phones are tossed in dumps every year, where their lead, cadmium, and mercury leak into the environment. That's monstrous.

In the video *Out of Yellowstone*, a film I watched at an environmental education center in Montana, a rancher and his wife raised children and cattle in Montana's rugged, expansive Centennial Valley just west of Yellowstone. The rancher said, "The dollar begins in the earth; after that, it's all just traded." He recognized that there is no other place to "get stuff" and to make a living. Although there is much in nature to which we don't attach a dollar value (for starters, air, daylight, precipitation, pollinators, photosynthesis), the rancher and I agree: nature is in absolutely everything you touch.

I toured my house to see its contents through a vibrant matter lens: my bed pillow, the nightstand, my coffee mug, tomatoes, the kitchen radio, paper everywhere, the wheelbarrow, my cell phone. These vital objects did affect and alter me, and I could envision bits of nature within, though in most cases, I was utterly ignorant about what earth elements comprised them, where those came from, and how they were transformed. And energy (utterly vibrant matter) was involved in each transformation and each delivery into my life. It sparked many questions I knew I would never answer: where did the oil to make the plastic come from, on what far-flung lands were the metals mined, and who manufactured them and how? In nature transformation comes distancing. And with distancing, comes a dramatic change in valuation. I value my cell phone more than the anonymous distant pieces of nature that formed it. How ironic that my phone shrunk and compromised the very wild nature I seem to value most—the same phone I use to take pictures of it.

Even if I succeed in seeing nature inside my stuff and I wholeheartedly agree that I and everything in my life are vibrant matter and part of nature, I do not always feel connected to the still-living nature in my regular, everyday life. There are days I arrive home on the bus and don't remember if I looked up at the Wasatch Mountains, or listened to birds, or noted the direction of the wind. It's hard to push aside the civilized city-ness and to connect the water from my faucet to mountain snowmelt, or to note yet another record temperature when I can just turn on the AC. Will I be able to know what is regular about the nature in my everyday life—and to recognize when it no longer is? Such disconnection is to my detriment.

꿍꿍

A survey about fifteen years ago found that 51 percent of the American public spent no time outside in a normal day (excluding walking from the house to car to office). Another 30 percent spent less than an hour outside each day. It is ironic that the purpose of these studies was to measure people's exposure to outdoor environmental pollutants—not exposure to sunshine and birdsong. These percentages are likely worse now with the steady march of technology into our lives.

For many, outside nature is an abstract environ. French sociologist and anthropologist Bruno Latour says the modern, urban self feels more and more removed from nature. Yet at the same time, the modern self is increasingly entangled with nature—from biotech and pharmaceuticals to a changing climate. The result, Latour concludes, is unresolved tension between everyday experiences that comingle us with nature, and with a cultural worldview that believes we control and direct nonhuman nature from afar.

Whatever the reason, the lack of time in and daily connection with nature is unhealthy. Nature is, quite literally, healing. In a famous 1984 study in *Science,* patients recovering from gallbladder surgery at a suburban Pennsylvania hospital whose hospital beds had a view of a leafy tree healed faster (and needed less pain medication) than those who saw a brick wall. "Just" a view of a tree.

Besides physical healing, time in nature—even wee bits of it—restores brain functions, such as ability to pay attention, perform various tasks, and be creative, all documented in a mounting number of scientific studies. Natural environments (meaning environments not dominated by human structures or pavement) are soothing because they allow for a gentle fascination and soft attention to objects and stimuli, in contrast to the sudden, switching, and demanding stimuli of our modern lives (ding, I just received a text). A key characteristic of healing nature is the ability to engage all of our senses, for which we need quiet places without cars, buildings, or other human trappings. *Scientific American* found that "Just three to five minutes spent looking at views dominated by trees, flowers, or water can begin to reduce anger, anxiety, and pain and to induce relaxation, according to various studies of healthy people that measured physiological changes in blood pressure, muscle tension, or heart and brain electrical activity." Imagine if city planners, schools, workplaces, and developers designed our living spaces with this in mind.

If getting outside is the first challenge for connecting with the everyday nature around us, the second is pulling back the cultural curtain, which is part of the aim of this book. We don't see nature as *it* is; we see nature as *we* are. It's why we believe there is an "away" to which we send our garbage and why a yard's shorn green grass expresses civic duty. The crescendo of city noise comes with a cultural belief that somehow we get used to it, even when it's collectively making us sick. People use the phrase "out of the woods" as though woods are dangerous, even though many believe this is where real nature lies. Some malls and restaurants are wrapped in culture, complete with "nature features" designed to make us feel good and buy more, well, nature.

Another challenge (for me and for most people) is to see *all* nature as connected and not partitioned into "wild" and "tame" boxes. At its core, the ecological crisis on this planet is a crisis of how I and my fellow humans use nature in our everyday lives. Those "all natural" snack chips involved industrial agriculture and palm oil deforestation. The water from the faucet represents a global hydrological system that is under great stress. My furnace symbolizes the heat of fracking deep in the earth in a faraway place. My laptop is a travelogue of metals and minerals mined in remote corners of the wild globe with cheap fossil fuels. How I consume and value these brought-to-you-by-nature products ulti-mately has more tangible impact on wild places than efforts to "save" wild places.

Everyday nature is where I engage the nonhuman most per-sonally and continuously, the elements of which are indivisible from nature's wilder cousin. It is impossible to separate the breath I just took from the air that jets and bees fly through from the air traveling inside the tree in my backyard or in the mountain trees beyond. Do I think of and value this "air" differently because of cultural wrapping and naïveté, or perhaps hubris and derangement? There is great value (indeed, urgency) in knowing and understanding living, vibrant nature and how it does what it does to produce both a towering spruce and a phone. And the best place to start truly knowing that nature is right out the front door.

Since all those Pokémon Go players are going outside, I hope that after they capture their monster, no matter if they are in Manhattan or Madison, that they pause, put down their phones, sit under a tree, perhaps spot a nonmythical creature, and watch the sun filter through the leaves as the clouds roll by. While gazing skyward, I hope they reimagine this grand yet ordinary nature—the beauty, the tensions and the consequences, and its fundamental connection to them. And I hope they return again and again, monster or not.

⊙ ⊙ ⊙

⊙ ⊙ ⊙

⊙ *1* ⊙

Living in a Circle
of Beating Hearts

It wasn't a good time to call with my request.

"You've gotta come see this!" I heard in the background. "Oh my god, look! He's climbing, oh, he's right on the porch, look, LOOK! This is so amazing. What great pictures."

Karen and Nancy were gushing about the latest photos on their "stealth cam" of a young black bear. I called to chat about the same bear—not to gush, but to ask them to remove their bear-level bird feeders. This bear had hit several area cabins and just had a run-in with my dog, which I hoped would prompt concern for their own dogs.

"Please, just bring the food inside until the bear moves on," I asked.

"Well, I don't know," Karen said slowly.

Karen and Nancy's cabin is a half-mile from mine—as the bear walks—in the high-elevation woods of western Wyoming. I summer here, far from Salt Lake City, where I teach during the school year. And my summers in the woods typify all the singular experiences and joys of watching wildlife: great gray owls hunting from the aspen, moose browsing at the meadow edge, warblers plucking insects from the branches. The encounters expand my human world into something much more brilliant and real.

There is good reason why people want to watch and get close to animals: of all the elements of the beyond-human world—geosphere, hydrosphere, atmosphere, and biosphere—it is the beating hearts in the biosphere to which we connect most deeply. As human ecologist Paul Shepard wrote, "Being human has always meant perceiving ourselves in a circle of animals." From prehistoric times to present ones, adventure among the Others (as he calls them) has remained central to our lives; our species emerged enacting, dreaming, and thinking animals. For reasons historical, spiritual, and biological, we cannot be fully human without them.

And, they are always physically near us—intimately so—even in cities. The animals with the beating hearts are the back-boned vertebrates—fish, reptiles, amphibians, birds, and mammals—though these vertebrates account for just 5 percent of all the creatures in the Kingdom Animalia. All Animalia down to smallest beetle or worm are motile (meaning they can move, at least during part of their lives), eat, and breathe (the Latin word *animalis* means "having breath"). Yet curiously, many "animal kingdom" websites make no mention of *Homo sapiens*. Our language reinforces this us-them boundary; when we say "animal" we typically mean the nonhuman ones. At least Shepard's term "Others" conveys that there are countless Other animals besides humans, but I prefer the word Kin to denote all of us.

Despite animals' unity in independent movement, appetite, and breath, and their interpenetration in our lives and culture, we engage in an ironic dance of distance with them. We seek the company of favored animals and search with camera and scope or mount the motion-triggered cam. Animals we label as "bad" are feared, shunned, banished, exterminated. "Wildlife" (like black bears) are perceived as bits of good and bad, depending on which side of our acceptable boundary they stand.

If *Ursus americanus* "belongs" anywhere, it is up here in the woods. But I learned in the summer-of-bears at my cabin that all my black bear experiences (mostly cultural) followed me all the way up here. For starters in childhood: Cinnamon, my beloved stuffed bear with a lullaby music box; TV shows like *Wild Kingdom* and *Grizzly Adams*; Winnie the Pooh books; a favorite book *Blueberries for Sal* (where a bear follows a berry-picking girl, slyly eating out of her pail); and the fact that I was a badge-wearing Smokey Bear Junior Ranger. No matter if I see a bear up here in the woods or one wanders into a suburban strip mall (which happened last year), I attach cultural meanings of "bear." This is just as true for a backyard squirrel or a mouse, and it affects not only how close we can (or want to) get to animals, but also how close we can *be*.

The first of several bear visits was my very first morning at the cabin in May. My golden retriever Maddie woke to a sound and planted her paws on the bedroom windowsill. I put on my glasses. A young black bear pushed the last of the birdseed into a pile with her paws and lapped it up with a dark tongue. The feeder, which I had hung just eleven hours before, sprawled on the ground.

I was proud of my feeder site: a lone, slender Douglas fir branch twelve feet off the ground, stretching almost four feet from the trunk. No bear in my decade of bird feeding had reached the feeder (some had tried), but this was "smarter than your average bear," and smaller—I guessed a two-year-old, light enough to shinny out the branch. The bear was also a repeat offender, judging from the blue plastic ear tags. I snapped some pictures through the window.

In an exercise in my environmental communication class, students list animals on two blackboards, one labeled "good" and one "bad." Quickly and with relatively little disagreement, they fill both boards. Then we examine what lies behind our gut reactions; what is it about charismatic megafauna that we like and about insects that we don't?

Based on extensive surveys, over twenty different factors influence whether we like or dislike certain animal species, such as size (we prefer large over small), intelligence, mode of locomotion (we like animals that walk instead of slither), and evolutionary closeness to humans (we like chimps). We prefer zoomorphism (animals who bond to humans), animals with individual personalities, and animals with relationships to human society (such as pets, farm animals, game, and exotic animals). We like animals with similar social habits, like parenting, pack loyalty, and helping each other. We are less magnanimous toward animals capable of harming us or our possessions, like crops or gardens. The students put black bear on the "good" board and grizzly bear on the "bad."

This list makes it clear: we like animals "like us," even if we are reluctant to put ourselves in the "animal" camp. This inhibits our ability to see who many animals are and all that we share physiologically with them.

A mother black bear poking her nose through the triangular vent window on our pale yellow 1960 station wagon on a family vacation was my first encounter with a bear. Her two cubs stood well behind her, watching. It was thrilling, and a little scary, but mostly thrilling. At Yellowstone National Park in those days, black bears were park panhandlers, looking for picnic baskets like Yogi Bear and Booboo (and the open Park dumps). To me, that mama bear was feeding her cubs, a good animal. She stood upright and used her front paws to take the marshmallows and bread slices offered through car windows (though our parents didn't let us feed her) and seemed friendly enough.

Whether bears or birds or bugs, animals are a key cultural metaphor we use to express core feelings and perceptions about the world and all our Kin. If an environmental group wants to "save" a forest from some threat, they focus on an animal who is an iconic symbol of that place. Sentient beings with beating hearts and a language of their own are symbolic barometers of our fundamental beliefs and valuations of nature—in ways that trees or rocks or rivers are not.

The bear left the grounded feeder and walked toward the bedroom window where Maddie and I stood. When she saw us, she stood, stared at us, and sniffed the air. Maddie gave a soft, low growl. I banged on the window and yelled, "go on! get outta here!" Nonchalantly, the bear resumed walking and circled the cabin, finishing at the bird feeder with a few more mouthfuls. Satisfied, the bear walked over to my car and put her front paws on the passenger door and sniffed (I later learned she smelled mice in there).

She then galumphed past the shed and disappeared over the rise. I made coffee and called several neighbors to be on the lookout.

~oqqeee

The late Stephen Kellert (and longtime professor at Yale University) famously described nine different attitude orientations toward animals based on a large national survey. Roughly a third of respondents were strongly Humanist (interest and strong affection, particularly for individual animals) and another third had Negativist-Neutralist attitudes (active or passive avoidance due to indifference, dislike, or fear). One-fifth of people were strongly Utilitarian (concern for practical and material value) and another fifth were Moralist (concern for right and wrong treatment of animals). The least common attitude orientations were Aesthetic, Naturalist, Ecologist, Scientist, and Dominionist.

Why are such attitudes important? These feelings guide our everyday portrayals and cultural communication about animals, and the destiny of many species depends on people's subjective feelings about them, whether mountain goats or monkeys, skunks or chipmunks.

We bond with wild and domestic animals, says animal ethologist Marc Bekoff, because they show us their feelings and hearts—both universal emotions like happiness, fear, anger, and sadness, and secondary emotions like sympathy, guilt, and jealousy. Animals talk with their tails, postures, gaits, gestures, mouths, eyes, and noses. Several species, including elephants and magpies, have mourning ceremonies for fallen comrades. Bekoff told of two grizzly cubs whose mother was shot, and the female cub cared for her wounded sibling, catching fish for him and waiting for him as he limped along.

~oqqeee

Five days later, Maddie and I awoke to clanking on the porch. The same blue-tagged bear pawed at the moose antler on top of the barbecue grill. On a walk the day before, Maddie had raced past me, clutching the heavy antler prize tightly in her jaw for the half-mile back to the cabin. When she let me hold it—a good-sized, palmated paddle—I discovered how she discovered it: it reeked, probably from some critter's scent marking. So I parked the paddle on the grill, intending to wash it the next day. Guess the bear thought it reeked, too. I banged on the glass and the bear startled and ran to the dead-end portion of the porch, high above the ground. She sat cowered in the corner, head tucked in, dribbling a little urine, looking scared and staring right at me. Poor thing. I waited until the bear rose to all fours, then I banged again and she galloped off the porch, down the meadow, and into the woods.

Black bears have come to cabins since people have built cabins, attracted by all manner of smells, provisions, and plantings. Bears emerge from their dens and let their appetites walk them through the seasons, first dining on winter-killed carrion, new grass, and aspen catkins. Next come grubs, rodents, dandelions and streamside frogs, snails, and fish. By late summer, berries are dessert. I once sat uphill from a black bear for an hour as it dexterously ripped apart a large decaying pine log and delicately lapped up all the beetles, grubs, and ants inside. Black bears, like many species (including us), are opportunists with an adaptable omnivorous diet. Their foraging patterns easily divert for "anthropogenic attractants," such as birdseed or garbage.

The attributes of a species—like the food-driven curiosity of black bears—are just part of what influences our attitudes toward them. Our attributes as human observers (such as education level, sex, income, residence, and experience with animals) also play a part. Women and urban dwellers have stronger affective perceptions than do men and rural residents. Children who have relationships with animals have positive emotions about animals and focus less on utilitarian value. For both kids and adults, direct, participatory contact with animals, such as bird-watching or hunting, is tied to greater appreciation, concern, and knowledge.

It is hard to generalize what people know about animals, except to say, it isn't much. Overall, we know most about pets, animals who harm people, and wildlife involved in emotional public issues (think wolves and grizzlies). The people least knowledgeable about animals live in cities of more than a million residents. Knowledge of wild animals in the United States is pretty abstract and indirect. For many, the generic *bear* or *trout* is the common level of recognition; a specific species reference, *black bear* or *cutthroat trout*, suggests greater cultural importance. Rarer still is when a person recognizes an individual of a species, like a particular bear.

So where do our cultural cues and animal facts come from? Folklorists say conversation and oral stories, children's literature, popular and commercial culture, public performances, elite culture, and scientific discourse. Pop culture and entertainment affect perceptions enormously, from Animal Planet to PBS. On a class field trip to the zoo, where my students listen to how people talk about the animals, we hear kids call them by the Disney character they played. A mom told her toddler who got excited by the "big kitty" cougar that "no, honey, that's a *bad* kitty." (Sadly, studies report that most people learn virtually nothing about animals at the zoo, save for how animals behave in captivity.) Regardless of where we live, we see more animal portrayals in everyday culture (advertising characters like the insurance gecko, animal sports mascots, eagles on T-shirts, news stories about cougars, YouTube videos, photos shared on Facebook) than we ever personally experience.

Cultural portrayals rely on a stereotyped shorthand to represent a particular animal—sharks are manic killers, swans are regal, ladybugs are cute—that convey every fear, admiration, envy, and longing we possess. In just seconds, shared meanings (even ignorant ones) of jaguars or pandas are exaggerated and exploited to sell you products. It's difficult (perhaps impossible) to parse the human-generated symbol of a particular animal from the living and breathing animal symbolized. In essence, what we see when we look at animals is our reflection.

We do not think twice about stereotyping an entire animal species, yet work to avoid stereotypes within the human species—Mexicans, gays, the homeless. Any stereotype makes it easier to marginalize, hate, and fear an entire group and see it as a threat to the established order. We stereotype that cougars or bears simply are a certain way and possess identical traits, in part because it's unlikely that we can differentiate individuals and their behaviors. Thus, when one cougar attacks a jogger, fear and loathing applies to the entire species.

Conversations and stories shape our beliefs of what animals are like. In our speech, animals are both verbs and nouns representing countless human conditions. As Shepard notes,

We duck our heads, crane our necks, clam up, crab at one another, carp, rat, crow, or grouse vocally. We. . .lionize and fawn. . .We fish for compliments, hog what should be shared, wolf it down, skunk others in total defeat, and hawk our wares. We outfox and buffalo those whom we dupe; we bug and badger in harassment. We hound or dog in pursuit, bear our burdens, lark and horse around in frolic. We bull, ram, or worm our way, monkey with things, weasel, and chicken out. We know loan sharks, possum players, and bull-shitters.

And, we see ourselves in animals' experiences. My friend Flo came to my cabin one Sunday to watch from my deck a nest of soon-to-fledge woodpeckers. The red-naped sapsucker parents arrived with food in quick intervals, prompted by the continuous hunger-begging peeps of the chicks. Between feedings, the chicks poked their heads out the nest-hole, surveying sky and earth. Suddenly, one chick clambered out and walked up the aspen trunk, swiveling its neck to see the world and testing its beak on the bark with a few quick pecks. In a couple of minutes it flew away, perfectly, on untested wings and landed in a fir downslope. Shortly, the second chick repeated the fledge. The male returned and searched for the chicks, poking his head in the hole, walking around the trunk and back to the hole, where he clung for several more minutes. Wow, empty nest syndrome. Flo, whose adult daughter had flown back East that morning, said, "I know how the adult sapsuckers feel. No matter how old they are, I always hate to see my kids leave."

Our attitudes toward animals are not merely the result of photos, plush toys, and beer commercials but of ancient influences. Biologist E. O. Wilson claims that our evolutionary heritage created biophilia (the innate tendency to connect with other forms of life) as well as evolution-derived fears and prejudices toward certain animals. Over millennia, our brains and bodies evolved through participation in the animal world as both prey and predator, thus animals are justifiably at the heart of human symbolism. For many Paleolithic peoples, it was the bear who became a model of ourselves.

Many attribute a grand turning point in our relationship with animals to Charles Darwin, who taught us that differences among species were largely differences in degree, not differences in kind. Darwin overturned the presumption that there was a special origin for humankind; instead, our species is corporally related (directly or indirectly) through millions of evolutionary webs to *every* other organism that we encounter. Recent science has also challenged the fixed boundaries between humans and nonhuman animals in terms of physiology, brain functions, and "human" attributes like consciousness, self-awareness, and emotion. In *Beyond Words: What Animals Think and Feel*, ecologist Carl Safina reminds us that consciousness seems necessary for all creatures who must judge things, plan, and make decisions. Octopuses use tools and solve problems as skillfully as do most apes. Honeybee brains have the same "thrill-seeker" hormone that is in human brains. And elephants live in matriarchal families built on close bonds and communication.

Three ponderous, moist piles of bear poop lay in the middle of the game trail. Maddie and I were halfway down the meadow below the cabin commencing a hike. The blue-tagged bear had not returned in three weeks, but a grizzly was reported recently near the highway about seven miles away. I turned around and headed back to the cabin for bear spray.

The phone rang; it was my friend Sara. Outside, Maddie barked excitedly. Since she is not a barker, I figured someone was coming up the driveway, but all I saw from the window was Maddie trotting up the drive, looking behind her. Sara and I kept talking. Maddie's paws scratched at the front door; I went to let her in, but she wasn't there. I returned to the back door and saw at the far side of the driveway a bear standing with her front paws on a tree trunk, but no Maddie. "I gotta go, there's a bear!" and threw down the phone. Then I saw Maddie by the woodpile, ten yards from the bear. I called her and she came flying up the back stoop, glancing back. The blue-tagged bear ambled past the woodpile, past the fir that once held the feeder and toward the stoop. When her paws thumped on the steps, I slammed the door several times, yelled, and pounded the glass. The bear showed no sign of fear or aggression. She padded back down the steps and strolled over the ridge. I do not know if the bear chased Maddie or if Maddie chased (or was ready to chase) the bear. My heart pounded. In the intensity of the moment, it didn't occur to me to use the bear spray.

Zach from Game and Fish asked if I knew the number on the bear's blue tag. I enlarged my photos and found it. She was a two-year-old female who was trapped as a one-year-old last summer after causing problems several miles west at my friend Phyllis's cabin. The bear was released in the Upper Green River valley—a couple of hours away as the human drives, or thirty miles of mountainous terrain as the bear walks. Phyllis fed all sorts of critters for decades, winter and summer, and no one could convince her not to. Phyllis died in March and left her property and cabin to animal welfare groups. And now her bear was visiting me.

When I learned the bear's story, I felt differently about her, this young girl bear I decided to call "Phyllis." In her short life, Phyllis had learned that cabins are sources of food, and that people mostly stay inside those cabins and don't hurt you. Many one-year-old cubs hibernate with their mothers a second year, but Phyllis survived on her own in the wild Upper Green. Amazing. No wonder she trekked back to familiar ground. Zach said the sooner a cabin-visiting bear is relocated, the better chance it has of "reforming." It was not good news that Phyllis was back.

Until I called Game and Fish, Phyllis was doing what she did every day of her normal bear life: sleeping, roaming, eating, pooping—nothing newsworthy by human standards. Wildlife "make the news" when a human makes claims about them, or when bound-aries—symbolic or real—between humans and other animals are blurred or are breached. Moose visits neighborhood, insects visit crops, duck nests downtown, bird collides with plane, black bear has run-in with a golden retriever. That is important to remember about "news" versus the long, languid stretch of "normal" in the everyday lives of animals. Some zoologists estimate that many animals basically rest and "do nothing" 80-90 percent of the time.

Government wildlife officials, like Zach, are the most predominant voice for animals when they make the news. Years ago, I analyzed news coverage of wildlife in four urban and four rural Midwest newspapers. About 60 percent of all quoted news sources were government employees, and 80 percent of all stories were about game animals. This reflects the dominance of game animals in wildlife management, the role of state wildlife departments, and the big animals we prefer; it also parallels where wildlife management dollars are spent. Urban newspaper stories focused on animal "stewardship," which makes sense in cities where people see wildlife infrequently or as a by-product of outdoor leisure. The predominant story theme in rural newspapers was "utilitarian," which makes sense in a place where people live physically closer to wildlife, and directly interact with them, and where many jobs depend on working outside. Rural Americans greatly value animals (wild and domestic) for what they provide, and dislike them for what they take or harm.

⁂

The day after Maddie's encounter, Zach towed a bear trap up the driveway. While we talked, Maddie nosed cautiously toward the round metal tube, then sprang back with fright, which she repeated several times. "She smells 'em," Zach said. He positioned the trap halfway down the long drive, out of view of the cabin. He baited it with two pieces of fresh trout and grain pellets, and mounded fresh dirt below the opening to catch paw prints.

The trap stayed for four days, but the only creature caught was a golden retriever who could not resist the fish. Zach replaced the fish and reset the door. When Zach came to remove the trap, the fish was once again missing, but there were no paw prints in the dirt. "Probably ravens," he said. Clever ravens. The day after Zach pulled the trap, Phyllis appeared on Bill and Jane's porch, neighbors a half-mile east. Jane was upset she didn't get pictures.

There are three choices for solving human-bear conflicts: try to change the bear's behavior (trap and relocate), kill the bear, or change the humans' behavior. Changing the bear sounds good, but early intervention for Phyllis did not work (and she likely was now trap-wary). Changing the humans is even more problematic. Karen and Nancy would not alter their big array of bird-feeding stations, nor would a dozen other neighbors. A couple of neighbors fed their cats outside. Zach once spoke to our neighborhood meeting and posted roadway signs about removing food sources when bears are present, but little changed.

My neighbors' sentiments about bears matched those reported in bear-human management studies. Most people like black bears, enjoy seeing them, and appreciate their place in the ecosystem. People perceive them as highly intelligent and aesthetically appealing. But sentiments are mixed about the threat bears pose and what "management actions" are appropriate. One neighbor (who cooked her bacon outside and didn't secure her garbage) was ready to shoot any bear on sight. Some neighbors warned me not to call Game and Fish for fear that Phyllis would just be shot. And a great many continued to refill bird feeders, either unaware of the phrase "a fed bear is a dead bear," or oblivious to their connection to it.

When mammals like black bears, foxes, coyotes, deer and others cross human boundaries (invisible to their eyes, of course), they are perceived as enigmatic wild creatures, or as destructive intruders. The boundaries are firm in urban areas—the "intruder" must go—but seem more fluid in a rural area, depending on the species (coyotes are often dispatched as "varmints") and perceived harm to livelihood. A bear's "trespass" is mitigated by our ecological concern for them and by their pervasive cultural presence. At my Summer Solstice party several days after the bear trap rolled away, Jennifer brought me a straw ornament of a bear as a hostess gift. I hung it near the kitchen sink.

The decor of most Wyoming cabins includes multiple representations of woodsy animals: bears on kitchen towels, antler coat racks, moose light-switch plates, couch throws adorned with trout, bear toothbrush holders, owl clocks, and artwork and photographs with these animals. The tchotchkes often repel me— the Disneyfication of a woods full of far more than charismatic megafauna and game animals. How tawdry using animals to hock consumer goods, the very production of which shrinks habitat and pollutes and warms our world. Why depict cartoonish moose standing on two legs? And a dopey-looking bear holding the toilet paper—really?

Yet, I haul home moose paddles, elk antlers, jawbones, and feathers. Egg shells, fossils, and assorted teeth line my bookshelves. It's the naturalist in me, my attempt to touch corporeal remnants of my Kin and feel I know them. In a way, these objects are as much collective markers of a specific place, a habitat, as the manufactured knickknacks—a way to feel I possess these woods and the Kin who live in them more fully. Both types of fetishes may contribute to an exaggerated (even false) sense of intimacy, that we somehow "know" bears or owls or moose. It speaks to the hunger we have for animals, the desire to have them near, whether actual fragments or humanized versions of them. We are "species lonely," writes author Thomas McGuane in *Some Horses* and therefore rely on needy dogs and cats to connect us to Earth's other inhabitants.

In contrast to other animals' rootedness in place, modern humans are adrift in standardized spaces and not solidly oriented to landscape or native creatures. While driving across Idaho recently, I exited to get gas and lunch, and for a moment I had no idea where I was because the franchises, billboards, and parking lots were the same ones I saw everywhere, anywhere. Other animals completely possess their home places with ease and confidence of movement and deep knowledge. Animal Kin live lives congruent with and rooted in their surroundings with skills and abilities perfected honed to it.

Many of us spend our lifetimes trying to get close to animals, in hopes that our meetings will magically bridge the chasm between their world and ours—as if these are somehow two separate places and not one in the same.

~oeeeee

As the summer blossomed and deepened, I missed the birds. I caught glimpses in the aspens or heard them sing in the morning, but without the feeder I rarely saw some species. Then I spied in a closet a small plexiglass feeder with suction cups to hang on a window, a gift I never used. Ah-ha, I could attach it to the office window that was twelve feet off the ground and above the cement walk-out basement; there was nothing for a bear to hold onto and climb.

Immediately the evening grosbeaks materialized, as did the Cassin's finches, black-headed grosbeaks, pine grosbeaks, and pine siskins. And what a view I had from my desk in front of the window, no binoculars necessary! Their slender talons clutched the feeder rim, their beaks frenetically opened the shells, their heads cocked when I spoke to them. I witnessed pecking order; the female evening grosbeak ruled the plexiglass roost, and the tiny nuthatches waited. I was a voyeur with a box seat on the bird world.

There is magic in the wild and close encounter that is hard to name or replicate. The best nature documentary is incomparable to seeing cow elk dance in my meadow, or startling a sandhill crane when I hike. It is like the difference between seeing a mountain peak in a postcard versus hiking there yourself; with all the exertion, anticipation, and engaged senses, you see and feel the mountain in full color and context. With other animals, it's powerful, intoxicating even, to be in their presence, to know you're sharing the air, the light, the moment. The magic is also the serendipity and uncertainty of the encounter, and even the possibility of danger. Animal Kin define and create the moment as much as I.

Like most photographers, I often snap a photo when the animal looks my way. It's seductive to think they returned my gaze, that we shared a moment of mutual recognition and they saw me like I saw them. However, when the hooved animals grazing my meadow early and late look up at me on the cabin deck, it's to judge my threat; it's a look of alert caution, nothing more. Their interest is eating (like the bear's) and I am judged as impeding that or not. They may recognize me as a human animal, a species that mostly means danger to them. If I want to watch them, I must keep my distance.

Yet on another level, our glances and interactions are interspecies communication with shared meanings and understood signals. Once many years ago, the communication was the purest I have ever experienced with another creature: I talked with a cougar. She had been watching me as I meditated on a rock at the edge of a cliff in southeastern Utah. When I stood and saw her, she took several paces away, her head turned back to watch me, the tip of her tail flicking side to side. I started talking to her, amazed how beautiful and regal she was. I took pictures. Then she turned to face me, sat, and wrapped her tail around her body. She gave me the slow, deliberate blinks of a contented cat. I believe we understood the other's intent at a palpable and visceral level; we were partaking of each other. It was reciprocal, deep, undeniably—unintentionally—honest. It was prayer, which in its most elemental sense is speaking *to* things, rather than merely *about* them.

Sure signs of spring for me are the smell of warm wind, gathering light, robins sparring in the backyard—and, when the nest cams go live. Wildlife cams around the world beam animal intimacies: puffins, snowy owls, penguins, bald eagles, pandas, and the fishing brown bears of Katmai National Park. One cam site's founder told Associated Press, "What we are doing is building out the zoos of the future, where animals run wild and people from everywhere can feel connected to the experience."

The animals are wild and the video feed unedited, which gives the viewer critter-life in real time with long expanses of normal quietude in their everyday lives. Hanging out. Sleeping. Flicking away flies. Looking around. Grooming. More sleeping. Even for fishing bears, there is lots of standing around in the water, watching, waiting, interspersed with occasional tiffs over riverbed turf. After enough cam viewing, a viewer learns what is involved in feeding chicks, catching rodents, and fishing for salmon.

In today's culture, we believe electronic encounters make us more connected and shrink the world. But as philosopher David Abram argues, the primary lesson you learn watching nature programs "is that nature is something you look at, not something you are *in* and *of*." The cams do allow me to enter the bear's watery restaurant as a silent voyeur half a world away. But there is no investment on my part, no consequences. I'm not sitting on the bank swarmed by mosquitoes, I cannot smell the rotting fish, and I'm in no danger from these talented carnivores. Do I have a stake in any outcome—theirs or mine? I'm entertained, I'm awestruck, but am I "connected"?

Compared to watching bears fish at the Anan Creek Preserve near Wrangell, Alaska—no. I took my dad on a small-boat cruise of the Inside Passage, and Anan was an optional day-trip. We were instructed to leave anything that smelled (food, gum, suntan lotion, soda) on the boat. We marched single file behind a burly woman with a rifle—no stopping!—along a wooden boardwalk to an enclosure of sorts. A series of wooden decks with railings overlooked a stretch of Anan Creek where an ensemble of black bears and brown bears fished for migrating salmon. No loud talking, no camera flashes. Sign up for fifteen minutes down in the blind at the river's edge. If you need to pee, a guard with a rifle will escort you to a porta-potty; open the door and wave when you want to be escorted back.

For four hours, we were bewitched by bears. Bears staked out fishing holes—black bears upstream, brown bears down. Cubs cavorted and climbed trees and napped in them. The smell of broken salmon wafted up to us, along with the rush of rapids where the bears fished. Above the rapids in a stretch of calm, shallow water, exhausted salmon who had run the gauntlet prepared to spawn. We identified bears with great fishing skills and those who missed a lot. Some bears walked their fish up the bank to eat the guts and brains, abandoning the rest for the long line of scavengers. Some bears ate their catch on a rock in the river before turning back for more. When a bear ambled by close (the guards on alert), I saw duff in its fur, smelled its dusky body. Though we were just yards from them, we felt invisible—that's how focused they were on the feast. But I realize that our presence likely changed the scene itself in some way. And I recognize that my Anan experience was one that only a privileged few ever experience.

Four decades ago, radio telemetry revolutionized wildlife biology, allowing biologists to attach small pieces of electronic equipment to a captured animal and track its location once released. Oh, the reams of data collected: cougars roaming great distances and birds migrating vast oceans. Biologists said it minimized animal disturbance and slashed hours in the field. Then came web cams, which ramped up the viewing several notches. Then came the drones.

One May, some tourists buzzed a herd of bighorn sheep in Zion National Park with a drone, causing young sheep to separate from the herd. The park ranger said they encountered people using drones in the park several times a week to get close to wildlife. Other national parks reported visitors disturbing nesting birds with drones. The following month, the Park Service prohibited drones in all national parks because the devices annoy visitors, harass wildlife, and threaten safety. National Park Service director Jonathan Jarvis said they would educate visitors about how drones disturb wildlife. (Just like managers educate cabin owners about bears, I wondered?)

Wildlife biologists argue that drones are once again revolutionizing their work, and they "need" drones to learn and understand animals in order to better protect and manage them. It is a typical anthropocentric perspective: animals need us and our knowledge, and any drone disturbance is therefore justified in the name of the larger animal (if not public) good. I trust that biologists are more sensitive about disturbing animals with "conservation drones." But is that even possible?

To a wild creature, drones in anyone's hands are paparazzi, an alarming buzz of unwelcome attention. It reduces the vital, private space animals need to thrive, and reduces the space doubly with the noise they bring. Like electronic dogs hounding a fox. Animals need quiet privacy and deep retreat; they know this and they seek it. And if we recognized that we are one animal among many animals (not the other way around), we would know that we need quiet privacy and deep retreat, too. But society is hell-bent in the other direction, ever more oblivious to privacy: we give it away on Facebook and phones that track us and the Google Earth camera that snaps close-ups of our homes. People "retreat" into ear buds and constant e-connections, which is not retreat, deep or otherwise.

The wrong question is, "How close can or should we *get?*" to animals through any means, drones or otherwise. The deeper question is "How close can we *be?*" in a way that is irrespective of distance. We live in a circle of animals, one community of countless beating hearts, whose fates are far more intertwined and synchronous than we can fathom.

⌇⌇⌇⌇

Though I did not see Phyllis again that summer, I have no doubt that she (and other bears) traipsed across the area all summer. Signs let me know they were near: scat in the woods, a paw print in mud outside the basement door, another print in the road dust. By August, bears smashed through the raspberry bushes a mile up the road, depositing crimson piles of evidence nearby.

When neighbors came together, we told stories of bears, stories that reminded us that Kin were fully present and proximal in our lives. Gail's bear tore down a bird feeder. Anne's dog chased a cub off their deck one morning. Tom got pictures of a bear on his cell phone, though he wasn't sure whether it was "my bear." Karen and Nancy got pictures of a bear visiting their porch on their stealth cam eight times. Only after the last visit, when their dogs cornered the bear and a shotgun blast in the air was necessary to end the confrontation, did they bring all their bird food inside.

But I was uncomfortable with our stories because the bears remained objects of our human gaze, a bit like circus bears who came to our cabins to perform for us, but sometimes crossed the line from desirable to unwanted, from close to too-close, from rewarding to costly. We were turning them into panhandlers, threatening the very creatures we so enjoyed.

Humans seem to feel safest with boundaries: inside/outside, private/public, wild/tame, nature/culture. Yet any such division demarcates two separate worlds—when there is only one world available. And we long to be in that same world with our Kin, a longing that is a deep-rooted piece of who we are, as imbued in our language and dwellings as in our hearts.

In 2016, an extensive study by the World Wildlife Fund found that between 1970 and 2012 populations of mammals, birds, reptiles, amphibians, and fish around the globe dropped 58 percent. I read the study several times, trying to fathom the unfathomable numbers. The report said biodiversity loss is much higher in low-income countries, and it correlates with increasing resource use by high-income countries. It's related to consumption, climate change, scarce water, and degraded land. In just forty years, almost 40 percent of terrestrial wildlife is gone, vanished.

Writer and poet Alison Hawthorne Deming ponders in her book *Zoologies* what the disappearance of animals means for the human imagination and spirit. She wants "to remember what it feels like to be embedded in the family of animals, to see the ways animals inhabit and limn our lives, entering our days and nights, unannounced and essential." What we need is a cosmology that is able to see ourselves as just one species among Kin in a global circle of circulating hearts—whether we encounter them in the backyard or the distant wild. A cosmology that holds as its general universal law that Kin are as elemental to our being as gravity, as essential to our biosphere as oxygen, as crucial to our hearts as the blood beating in them.

In *The Others*, Shepard composed a letter of reply from the animals to the humans, which concluded:

> We are marginalized, trivialized. We have sunk to being objects, commodities, possessions. We remain meat and hides, but only as a due and not as sacred gifts. They have forgotten how to learn the future from us. . . . Their own numbers leave little room for us, and in this is their great

misunderstanding. They are wrong about our departure, thinking it to be part of their progress instead of their emptying. When we have gone they will not know who they are. Supposing themselves to be the purpose of it all, purpose will elude them.

One late summer evening after a sudden heavy cloudburst, a black bear lumbered up the steep meadow. The storm quashed my BBQ plans and I cooked veggie burgers inside; when the pan got too hot and smoked, I opened a window. The bear was large and rotund, far larger than Phyllis and without ear tags. He rounded the big Douglas fir twenty yards below the kitchen door and then stood, sniffing the air, his massive paws and pencil-thick claws resting on his belly below his beating heart and a white chest blaze. He fell to all fours and took a dozen more paces up the meadow, stood again, and sniffed again. I grinned, agape, clutching a dish towel. The bear turned and in a few large strides, cleared the ridge and disappeared. Thump, thump, thump went my heart.

The Granddaddy
of All Trash Days

A trash pile the size of two cars yawned into the street. At its base lay chunks of concrete and dirt clods stacked several feet high. Lying limply over the concrete was a mattress, blue flowers on a once-satiny white, ripped to reveal an underbelly of springs. Topping it was a roll of olive green carpet, a closet door with a hole in it, and a tangled mound of chain-link fencing. Next sprawled several large limbs of a Siberian elm, a vagabond volunteer tree that escapes a mower's blades along fence lines and foundations. Its dead leaves rained onto the layers below. Perched precariously at the summit were an assortment of cardboard boxes, a garden hose, several tires, and plastic resin chairs—two white, one green.

I have been watching this pile across the street grow, waiting for trash day—not just any trash day, but the granddaddy of all trash days. Once a year in Salt Lake City, crews pick up anything you stack at the curb (well, almost anything: not hazardous waste like paint or pesticides, not tires, and not batteries). Each neighborhood is assigned a weeklong pick-up slot sometime between April and September, and ours is next week. As the piles expand, cars compete for curbside space and maneuver past piles breaching into the street.

I am a voyeur of these garbage piles on my daily dog walks. Unlike the waste hidden in bins and removed by a trio of trucks each week, this trash is piled in the street for the world to see.

It's a fascinating window on what wears out its physical—or just desired—life span. But garbage reveals far more than what we discard; it tells us how we value the things that fill our homes and lives that we use to relate to each other and the world. It tells us how we value things of the Earth. Rather than ignoring it, it's instructive to examine "garbage" more closely.

After a long dog walk, I sat in the rocker on my front porch, drank a soda, and watched birds in the tall pine across the street with my binoculars. All of these objects—rocker, soda can, binocs— were pieces of nature. The rug the dog lay on was woven from recycled plastic bags, made from petroleum. The rocker was wood from somewhere. The aluminum was probably mined in Australia or Brazil. The glass binocular lenses were made from sand, soda, and lime; the outside casing was a hard molded plastic; and the strap was woven, probably from synthetic fiber. By the time I brought this stuff home, it was so transformed by manufacturing—and the water, energy, and machines it requires—it was easy to forget it was stuff of the earth. Absolutely everything I touch in my daily life—coffee grinder, toothbrush, laptop—was built with elements from nature. Man-made is more appropriately called nature-made.

I walked to the side of the house and tossed my soda can in the bin. At the fleeting moment when my fingers released it, clink, plop, it became trash—waste, garbage, rubbish, refuse—and another thing entirely. In the instant when I physically separated from it, I also psychologically separated from it, even though the can had changed not a whit from the moment before. It was desired and valuable; then, it was unwanted and valueless. I was hardly conscious of the change, yet it is a powerful transformation during my frequent and often brief acts of consumption.

This is our culture's relationship with garbage: it's the end point in a simple linear process—produce, consume, dispose—and hauling it away is the divorce, the final separation of these products from our lives. Except, it's not—not simple, not linear, not separation. When I stare at these trash mounds, what I see is life and energy and the richness of nature. I see priceless elements with countless possibilities for our living with the earth in a cycle of birth, death, and return. The time to think about garbage is long *before* we throw it out.

ⱺⱺⱥⱥ

For several springs after we bought our respective houses, my friend Maureen and I journeyed to the landfill to get wood chips to cover and retain moisture in our flower beds. We stabbed pitchforks into the towering pile, filling a metal garbage can. The deeper into the chip pile we dug, the more it steamed in the sun, the more it smelled like the woods after a rain. The chips cost $10 for as many bags as we could cram into the car, usually a dozen or more.

The wood was collected by city crews: Christmas trees, dead trees, old wooden fences, fallen limbs. Next to the wood chip pile where we dug towered mounds of topsoil, made from yard waste collected from the tan yard-waste bins provided by the city free of charge. Wormy apples, coffee grounds, grass, carrot peels, leaves, and weeds—all decomposed and transformed to humus, the organic matter essential to the fertility of the earth. Humus is the rich, fertile loam from which all life begins and all life will eventually return (except plastic: more on that later).

Yet very little waste is ever returned to the earth as nutrients and humus. Organic waste of all kinds (food scraps, yard waste, wood) is the single largest component in our trash (34 percent). By volume, 20 percent of the landfill is discarded food: food damaged or abandoned in agricultural fields, blemished grocery produce, restaurant leftovers, and food from our fridges. One estimate of individual food waste is twenty pounds per person each month. Thus, your food waste is like buying five bags of groceries and leaving one bag in the parking lot.

Composting defies the traditional linear garbage cycle because it closes a circle and returns nutrients to the earth from whence they came. In a rather Zen-like way, nature knows no "waste" and has no beginning and end-points; nutrients and energy continuously cycle through birth, life, death, and decomposition in prairies, woodlands, and oceans.

In *The Unsettling of America*, Wendell Berry conceptualizes "return" as an energy community where all plant, animal, and human bodies are joined. The farmer-philosopher explained that life and death "are indissolubly linked in complex patterns of energy exchange. They die into each other's life, live into each other's death. They do not consume in the sense of using up. They do not produce waste. What they take in they change." Upon death, a body and its energy are returned to the earth, and through decomposition that energy becomes available for new life.

Berry explains that by working with nature rather than against it, agriculture can be sustained indefinitely and can operate as a complete cycling system that returns nutrients to the soil, aiding soil organisms and all that sprouts from it. A defining shortcoming of modern agriculture has been the abandonment of these cycles, which requires evermore inputs of energy, fertilizers, and chemicals to keep the processes moving and restarting.

Salt Lake is lucky to have free yard waste pickup, but participation is entirely voluntary and bins must be requested. Many cities around the country do not have any organic waste pickup; a few cities have gone to the max. At his house in Berkeley, my brother showed me how to separate the garbage into a slotted bin, with one slot for all compostable material—not just the apple core and grass clippings, but also the moldy cheese and chicken bones. The rule: If it was once alive or it's soil or food, then it is compostable. The Bay Area built "hot" composting facilities with temperatures high enough to kill bacteria in food scraps and decompose them. It is law that residents separate all their garbage, and it is enforced; teams inspect garbage bins and knock on doors to educate residents. Hot composting also removes the smelly and rotting parts from a landfill that produce methane gas, a very potent greenhouse gas.

⌇⌇⌇⌇⌇

Composting is the gold standard of cyclical return, but some practices merit a bronze: extending the useful life of objects destined for the waste bin and giving them second life through reuse and reformation.

Last week when walking the dog, I saw a woman deposit two chaise-lounge cushions on top of some cardboard boxes in her pile at the curb. The cushions looked practically new. They were gone the next day when we walked by. Salt Lake's annual curbside cleanup has a food chain all its own with trash scavengers, professional and amateur, cruising the streets to pluck items before cleanup crews arrive.

When my kitchen remodeling coincided with trash week, I built my first pile at the curb. The contractors said they could haul this stuff to the dump themselves, but what the heck, the city would do it for free. Overnight, the pile shrunk. First, the oversized 1917 kitchen sink disappeared. The next morning, a few cupboard doors went missing. Then a beat-up flatbed truck rolled by slowly, trolling for trash. The truck's flimsy wood sides, wired together, looked like a Depression-era vehicle stacked high with possessions: solid wood doors, a cast iron claw-foot tub, an electric stove, a bike, sheet metal, and window screens. The driver got out and poked at my pile with his boot. He looked at the truck, looked back at my broken tile countertop, and then called for his buddy to help lift it. Eventually all that remained in my pile were broken pieces of plaster and odd ends of trim molding; I scooped the remains into my garbage can.

I was grateful for the reuse. I grew up with Depression-era parents with a deeply ingrained ethic of repair and reuse. Weary clothes became work clothes, then rags. When the car ran over the hose couplings, Dad repaired it, as he did leaky faucets. Mom repaired clothing, darned socks, and reupholstered the couch.

Compared to the fifteen million people worldwide who survive by picking the reusable or recyclable from putrid garbage dumps, Salt Lake trash pickers pick through piles of scentless affluence. In some countries (notably China, India, and Brazil), 1 percent of the urban population works the streets or open dumps—society's most impoverished and vulnerable, including children, migrants, disabled, and elderly. Digging through the rotting discards of others provides their only source of income and/or food. Ironically, waste pickers in Surat and Salt Lake both demonstrate how woefully unable our waste systems are at finding second life for things on the other side of our fleeting desires.

My city is full of thrift stores and consignment shops, which I hope exists country-wide. We also have scrap metal collectors and a Habitat for Humanity ReStore with donated leftover building materials. On my neighborhood walks, I have seen deck chairs made from old skis and yard sculptures welded from all manner of car parts. The Spotted Door, an online store started by a Salt Lake man, sells up-cycled products made by crafts-people: drinking glasses from wine bottles, jewelry from bullet casings, and totes from rice bags, as well as furniture and toys. When I wanted to give away moving boxes, I posted an ad on the Freecycle Network, a grassroots volunteer nonprofit move-ment with five thousand groups worldwide trying to keep stuff out of landfills.

But for many people, this second-life network of reuse is invisi-ble, and civic information about repurposing is sparse. And second-use will forever struggle against the marketing mentality that shiny new is best. It also struggles against highly subsidized garbage services, against products so poorly made they expire before a second life, and against the dirt-cheap price of resource extraction and disposal.

My friend Camille once called me in tears. Developers were tearing down a house adjacent to her property, a peaceful enclave on the banks of Big Cottonwood Creek draining from the Wasatch Mountains. "They're just bulldozing everything and hauling it away!" she cried. In a few months, two brand-new gargantuan houses were built where one 1950s rambler once stood. Camille witnessed the razing of fifteen houses in her neighborhood, twelve of them for a parking lot. "It seems like such a farce recycling my newspapers and cans when people can just waste all those per-fectly good houses," she said. "It's like a very bad joke."

As bulldozers tore down old dormitories on campus just east of my office, I called to complain. I watched the wrecking ball career into wall after wall, followed by a tractor that scooped the buildings and all their contents into a long line of dump trucks: metal bed-frames, mattresses, curtains, glass, insulation, and dark red brick—all smashed together and carried to the dump. Campus officials told me they accepted the cheapest bid. "There's nothing we could do," one woman said. It too made all the campus recy-cling bins seem like a farce.

In my neighborhood, "tear-downs" of perfectly fine houses are now commonplace. One woman told me they decided not to expand and remodel the house they bought because their builder said a tear-down "was a lot cheaper." Cheaper only because the economic deck is stacked for waste and cheap extraction of virgin materials and against reuse.

A city-planning professor at my university describes old buildings as full of "embodied energy." In addition to physical materials like concrete, steel, and glass, great time and energy went into constructing them; hence, buildings are incarnate structures of value, worth preserving and refurbishing. Concrete is embodied (literally) with ancient and delicate sea creatures that floated in the ancient oceans, then died and feathered down to the sea floor and were pressed to form limestone. On the hills above the university sits the skeleton of an old lime-kiln, which heated limestone to great temperatures; the stone was ground to a powder, mixed with sand and pebbles, added to water, and poured. What an astounding and poignant journey: ancient slender sea creatures as the essence of something so strong and substantial.

I daydream about a day when second-life (and third- or fourth-) for goods of every description is the cultural norm. I imagine a city-sponsored, year-round swap meet in an old warehouse where people donate and shop with the same enthusiasm as participants in the granddaddy trash day. I envision bragging rights over repurposed hardwood and exclamations about a vintage dress and resoled shoes. And I eagerly await the day when our recycling bins are rarely full—their contents reused and diverted, or never used at all.

⌇

Across the street a few houses east, a thrift-store quality couch and chair are parked at the curb as though awaiting guests. I watched a man stack wooden fence slats next to the furniture (he recently replaced his paint-peeled fence with a new white plastic one). On top of the wood, he threw an old mop and armloads of clothes. Next, he balanced an assortment of cardboard boxes. A postcard from the city clearly instructs that wood be stacked separately from other trash and that cardboard boxes be put in the blue recycling bin—good rules, which I have never seen enforced. On pickup day, the wood-compost truck stands idle while the entire pile is scooped into the garbage truck.

When discussions in my classes concern environmental issues, someone invariably says, "But I recycle." Recycling feels like such a great free pass that lets you off the environmental hook. The bins are indeed seductive; they make you feel good, that you are living a green, politically correct life. Nowadays, more Americans recycle than vote. However, what seems like such a successful environmental behavior is seriously flawed—both as a solution to our disposal and as an argument for our innocence.

A friend once called me a "recycling curmudgeon." I am not; recycling programs have kept an enormous amount of municipal waste from landfills, and I am a dedicated recycler. But for five reasons, recycling is far more problematic than we are led to believe. First, not everyone plays; recycling is not available in many cities or it's voluntary, like it is here. Second, if outright disposal is the worst option, recycling is just one small step less bad. Both choices pale when compared to the best options of preventing, minimizing, and reusing in the first place—like water fountains and reusable water bottles instead of drinks in plastic containers. Third, "single stream" recycling where all materials are combined and then later (kinda) sorted is inefficient and produces a much lower quality material (like when white office paper is mixed with shiny colored ads and newspaper) that is far less valuable to recyclers. Up to 15 percent of the material in a mixed recycling bin eventually goes to the dump because it's unusable or contaminated. Fourth, recycling requires large quantities of brand new energy, water, and other materials; refilling beer bottles instead of melting them to manufacture new ones makes far more resource sense. Or, in the case of plastic bags, the only tenable answer is nonuse: recycling plastic bags costs four times what the raw petroleum materials are worth. Finally, recycling is still disposal and does not move us toward a circular process where waste is not created.

If I had a magic wand, the one product I would disappear would be bottled water. "But I recycle my bottle" is what I hear as the justification for buying bottled water, even though just one in ten bottles is actually recycled, and even though the purchase was unnecessary in the United States with great and essentially free tap water. The lure of convenience and advertising images of bottled water as healthier and safer (it is *not*, by the way) convince shoppers to buy it by the caseload. Promising to recycle one-time-use plastic bottles does not change their highly toxic manufacturing process or the fossil fuels from which they are made.

One spring, my colleague Danielle and I visited a large recycling sorting plant on the west side of the valley. I wanted to be able to picture how the jumbled mess I tossed in my big blue recycling bin was unjumbled and sorted. Wearing hard hats and earplugs, we followed our guide to the initial sorting line. Heaps of mixed items flew by on the conveyor at a dizzying pace, as four workers picked and grabbed unacceptable items and flung them aside—garden hoses, large wads of plastic sheeting, kitchen waste, rope, clothing, and items plucked too fast to identify. After watching the speeding trash for just a few minutes amidst deafening noise (even with earplugs), I felt nauseated.

In the vast building, an army of machines pulsed and clanked and rattled. Machines sorted aluminum from tin with an electrical impulse, machines separated items by weight, machines shot bursts of air to separate light plastic from heavy plastic, machines shone an optical beam to identify and whisk away clear plastic, and machines separated paper. There was so, so much paper. From an upper level, paper fluttered down to the main floor like perpetual confetti. Front loaders shoveled paper here and scooped cans there. We walked past shiny square blocks of crushed aluminum cans the size of cars, stacked floor to ceiling, ready to load onto railcars and semis. All the commotion and ruckus before us was just sorting and separating; actual recycling would take place later in factories far away.

The good-news story is aluminum, where recycling makes far more economic and ecological sense than making new. Aluminum cans are made from bauxite, which is strip-mined in Australia,

China, and Brazil. It is refined to a white powder, subjected to electric current to separate out the aluminum, melted and added to other metals, and rolled into large sheets. Until I researched this, I had no appreciation for the can's former connection to the earth. Aluminum cans are valuable enough that people wander roadways to collect them, yet the United States now recycles just half of the aluminum cans it produces. Glass also recycles well (though only one-fifth of it is); it can be recycled again and again with no loss in quality or purity—particularly when clear and green grass is separated from brown glass before crushing. The biggest obstacle is the economics: transporting heavy glass long distances to the nearest recycling plant and using lots of new energy to melt and reform it may tip the balance against glass recycling. Recycling paper is good news–bad news: cardboard and paperboard recycle well (when separated) as does white office paper. But recycling programs often collect all paper together, making a far less usable product.

The bad news story is plastic, which has a host of recycling limitations. As journalist Heather Rogers notes in *Gone Tomorrow: The Hidden Life of Garbage,* the recycling numbers put inside the chasing-arrows recycling symbol tell consumers that containers are recyclable and perhaps even made with recycled materials themselves; often, neither is entirely true. The numbers do not really create a sorting system that is very useful for producers (or recyclers) who categorize plastics based on how they are made. And, 20-30 percent of U.S. plastic recyclables are shipped to other countries, mostly in Asia. It is estimated that 50 percent of the plastic shipped overseas is so contaminated with incompatible materials (including plastic lids left on) that it cannot be recycled and is instead dumped, often in unlined, unmanaged sites.

The more I read and learned about plastics, the worse I felt. I passed through my house, determined to root it out and shun future purchases of it. The quest was an utter failure: my glasses, the bread bag, toothbrush, shampoo bottles, clock-radio, TV, much of my car, doormat, and—the ultimate irony—the garbage can itself. Plastic is handy and so ubiquitous: it's cheap to make (because the full cost of petroleum products is not counted) and can be formed into rigid CD cases or soft, pliable binkies.

Though you might envision a brand-new life for plastics after recycling, what is in store is a far lesser life. Of the seven recycling numbers stamped on plastic goods, five are rarely or never able to be recycled; numbers 1 and 2 are the most recycled. Plastics lose a great deal of their structural integrity during reprocessing (even with the addition of new plastic and polymer), eventually making them unusable for recycling, a consequence called down-cycling. Thus, the plastic flowerpot that was once detergent bottles cannot be recycled again. Recycling only delays, but cannot avoid, plastic's final disposal as trash.

Yet, the world is making and discarding ever more plastic—twice as much now as twenty years ago. A 2017 study in *Science Advances* used industry figures to calculate that since 1950, industry has made 9.1 billion tons of plastic, enough to bury Manhattan under a *two-mile-high* pile of plastic. And, except for a small percentage that was incinerated (which releases toxic gases) or is still in use, 5.5 billion tons of plastic is still with us, floating in oceans and waterways, littered on land, and lying in landfills.

On every street the dog and I walk, most every trash pile contains plastic: lawn furniture, plastic wading pools, toys, blue tarps, plastic tables, shelves, wastebaskets, crates. What makes plastic a problem as trash is that plastic is a forever-product; there are no natural processes able to break it down (at least within human lifetimes). That's right: none of the commonly used plastics biodegrade.

Many people have heard of the Texas-sized trash-patch in the Pacific Ocean; in reality, there are many gynormous swirling plastic conglomerations in oceans the world over. Ocean plastic kills marine mammals and birds and creatures on shore. Photographer Chris Jordan documented albatross chicks slowly dying after ingesting plastic washed up on their remote beach, thousands of miles from civilization. The chicks pecked and swallowed assorted plastic trash, believing them bright bits of food. After I viewed picture upon picture of a bird carcass whose gut was chock-full of plastic, tears streamed down my cheeks. It's likely we are swallowing these bits as well; oceans pummel our discarded plastic into ever tinier *microplastics*, which are swallowed by the sea creatures that we then swallow. There is no "away" to which we can banish our plastic; it's just a matter of how, when, and where it is returned to us.

Journalist Rogers calls recycling "the politics of containment" and documents its very close ties with the waste industry. When recycling gained momentum in the early 1970s, it was not threatening to the garbage industry because recycling did not reduce consumption or mandate reuse. Instead, the waste industry diversified and grew by hiring recyclers to manage this waste, ensuring continued production and new profits.

The night after I toured the recycling-sorting plant, I dreamed that I was standing in front of its massive doors, trying to push them shut but paper whizzed through the gap and plastic bottles zipped between my feet. As the trash pummeled me, I screamed, "You shouldn't be here, go home!"

A professional organizer concluded that the average U.S. household has 300,000 items. That may include pieces of paper but I get the point: our houses are crammed with stuff. The story of garbage is the story of how and what we consume.

An article in *Sunset* profiled a family of four—living in a modern, airy, and modest home—that had transitioned to "zero waste." They undertook monumental recycling, donating, and whittling down of their possessions. They bought used clothes, borrowed books from the library, and carried bulk groceries home in their own containers. The woman in the article said that she wanted to have a relationship with every item in her house, down to the carrot peeler.

This unusual family managed to remove itself from frenzied consumer society; advertisers could not trigger anxiety and desire to buy, and Black Friday meant nothing to them. They repositioned their lives around well-being instead of well-having. My simplicity efforts pale in comparison, though they are important and useful. I do not let convenience guide my purchases. I buy in the bulk aisle. I avoid precut vegetables in plastic containers. I repurpose the slim bag around the newspaper as a dog-poop bag.

I use cloth napkins and dishcloths instead of paper towels. I eat in restaurants that serve food on dishes, and when I have a party, I use my china and flatware.

But there are still plenty of single-use items I dispose: plastic wrapping on the cheese, the plastic shell that held blueberries, the envelope from a letter. A staggering 80 percent of our trash is comprised of items used *once*, then thrown away. And almost one dollar of every ten dollars that Americans spend on food and beverages pays for its packaging. That packaging (which created the third largest industry in the country) is basically the production of immediate, *premeditated* waste.

Wouldn't it be nice if the consumer mantra of fast and convenient gave way to Slow Consumption where we would show pride for things *not* bought, praise for items we had long and lasting relationships with, and cherishing of goods to which we gave a second life. Thoreau wrote that a person is rich in proportion to the number of things she can afford to let alone. During a class discussion one day about consumption, I confessed that I wear some clothes from thrift stores and passed along to me by friends. Several students said I should be proud of this, but one student said there was no way she would ever wear someone else's clothes—it was just too gross.

Yet, despite all that I or any one individual does to reduce consumption and disposal, individual efforts alone are woefully inadequate. In his essay *Forget Shorter Showers*, Derrick Jensen argues that a focus on individual action (such as water conservation) is a systematic misdirection when 90 percent of all water is used by agriculture and industry. Likewise with waste. The EPA says the average American produces about five pounds of waste a week. But that figure does not include all the waste produced during mining, manufacturing, and transportation, which raises the total to forty-five pounds per person per week. Jensen notes that municipal waste—all the waste that cities and their residents produce—accounts for a paltry 3 percent of total waste production in the United States. We should live simply and nonmaterialistically, he says, but we should not equate personal change with the substantial social change needed to change the industrial economy that is killing the planet.

These ratios are astounding. They rightly direct our attention to waste prevention and minimization by extractors and manufacturers. Although most consumers are not aware of all the waste tied to each product they buy, it's nevertheless our consumption around the world that drives the extraction and attendant waste. Short of large-scale activism calling for a redo of the industrial economy (which I don't see on the horizon), changing how products are designed and made would change our granddaddy trash mentality and better mimic nature's absence of waste.

Most of our garbage results from a model of take–make–waste. Thomas Eatmon, a professor of environmental science at Allegheny College, said that unlike nature, where all waste is actually food for something else, human extraction and manufacturing systems are highly linear. We extract raw materials like copper and clay from the environment and then use tremendous quantities of fossil fuels to move and process the materials for consumer products. As much as 95 percent of the original material extracted—trees, ore, fish—ends up as waste. When the consumer product (whether a cell phone or a toaster oven) is disposed in a landfill, the extraction starts all over. It's an enormous waste of useful material and energy, he said.

Eatmon reports that environmental scientists are developing new approaches to manufacturing and marketing consumer products that "design waste out of the system" by treating waste as a potential resource. Such approaches go by many names: zero waste, industrial ecology, extended product (or producer) responsibility, and others. Some call these approaches the next industrial revolution.

In a manufacturing cycle, if a product cannot be decomposed into nutrients (like carrots or chickens), the next best thing is for consumers to return products to the producer so they can be disassembled and the constituent parts reused in some fashion. Instead of cradle-to-grave, think cradle-to-cradle, where all materials are kept and used in continuous cycles, ideally with renewable energy.

Michael Dell of Dell Computers once admitted that their inventory is meant to have the shelf-life of lettuce. Short shelf-life means you soon buy another and pitch the "old" one. Over 500 million electronic products are tossed each year; only 12 percent are "recycled," which often means they are smashed or burned to extract just the gold and other precious elements.

It doesn't have to work this way. For their class project, a group of students in a mechanical-engineering class at Stanford University designed and built a fully recyclable laptop they called the Bloom. Not only is the Bloom made of 100 percent recyclable parts, but the average user can take it apart in two minutes without tools. Users can remove certain parts and replace them without chucking the whole computer. Because it's so modular, users can customize it to suit specific needs.

In a YouTube presentation, student Aaron Engel-Hall said that most laptop users are little involved in the end-of-life of their electronics (other than stashing it in a basement or tossing it in the trash), but the Bloom laptop involves users. When a part gives out, the user easily removes it and puts it in a prepaid envelope to get a replacement part. This is not possible in traditional laptops, where plastics and various metals are soldered together, making it nearly impossible to separate them for recycling. It was an impressive presentation; however, it's been eight years, and the Bloom has not bloomed its way into the marketplace.

But Europe has been taking action. In 2003, the European Union passed the Waste Electrical and Electronic Equipment (WEEE) Directive, which mandates extended producer responsibility. Essentially, manufacturers of electrical devices must take back all products they sell after the products' useful lives are up. To comply with the legislation, EEE producers "need to consider the entire life cycle of electrical and electronic products, including the product's durability, upgrading, repairability, disassembly, and the use of easily recycled materials." Even though it has loopholes, the law is nevertheless an important step because electronics recycling has been what Annie Leonard (author of *The Story of Stuff*) called ecologically filthy. If more countries required extended producer responsibility for laptops, cell phones, and tablets, it would have a major impact on the way many companies manufacture and market their products—not to mention the thousands of metals and earth materials mined and then discarded.

Last year, my Pentax binoculars fell off the bookcase and the lenses no longer focused. They were my first good pair, which I bought after graduate school twenty-five years ago. I searched all over Salt Lake and the Internet for someone to repair them. One man said I could try to send them directly to Pentax headquarters and gave me the address. I kissed them goodbye and boxed them up. A month later they returned, good as new, no charge. I was grateful the producer still took responsibility for its product.

The biggest argument against changes to producer responsibility is that manufacturers claim it costs too much and would increase prices for consumers. To me, it puts the durable back in durable goods and places the costs where they should be—on the takers and makers and users of the earth's materials. It considers the consumption of energy, raw materials, land, and water, not just the cheap subsidized trash at the other end. It would help us think of the Earth not as twenty-four-hour convenience store but as a bank whose wealth must be safeguarded.

～🦋～

A trash pile a few blocks from my house stretches the length of two houses. At one end are three big picture-tube TVs and a worn entertainment center with a tape deck and a CD player placed on its shelves. A couple of broken chairs, an end table, and a love seat are positioned next to the electronics like a street-side living room. The center of the long pile is an amalgamation of clothes, plastic, bent tomato cages, a Mr. Coffee, and a microwave. At the other end lie heaps of old carpet reflecting outmoded fashion choices—gold, avocado, ultra shag.

When Interface carpet founder Ray Anderson read Paul Hawken's *The Ecology of Commerce*, he wept. He realized what the industrial system was doing to the Earth and that his company was a traditional plunderer. This "spear to the heart" led him to transform his company to a business focused on authentic sustainability, using a cyclical model that mimicked nature and its closed loop. Beginning with design and raw materials, the company undertook lifecycle assessments to evaluate the impact of their carpet making on resource extraction, toxicity, and global warming. Key was "de-materialization," or making the same product with less material and/or using recycled and bio-based materials, which reduced new material extraction, energy use, emissions, transportation, and waste. For example, their assessment led to the use of recycled textiles for carpet backing.

Interface also innovated through biomimicry, using the ingenuity and design of nature as a model for solving human engineering challenges. Over billions of years of nature's R&D, its creatures, plants, and microbes have developed skills, features, and strategies that help them excel at survival. By asking how nature designs a floor, Interface developed modular carpet squares inspired by the "organized chaos" of the forest floor. The patterns and coloration of the tiles resulted in less manufacturing waste and allowed for nondirectional installation, reducing waste to 1.5 percent compared to as much as 14 percent for traditional carpet. Mergeable dye lots let customers remove and replace individual tiles without disrupting the overall design. Customers return worn-out carpet squares to Interface to be recycled into new products.

Inspired by geckos adhering to vertical surfaces, Interface developed TacTiles, small adhesive squares that allowed the carpet to "float" on the floor. The squares use no glue so the carpet is easier to replace and lasts longer. Interface said there was less waste, low off-gassing, and an environmental footprint that was 90 percent lighter than traditional glue adhesives.

A true closing-the-circle innovation at Interface is Net-works. This program enlists residents in seaside communities (one video showed the Philippines) to collect discarded fishing nets from oceans and beaches, nets that ensnare sea creatures and fish. Interface buys the nets from the villagers and recycles the highest filament into yarn for Interface carpet. Net-works cleans up oceans, protects sea creatures, provides valuable income, and closes a pernicious trash loop. This example of industrial ecology kinda makes me want to buy some carpet.

~gggee~

The city sent me a survey about the granddaddy trash day (which they call the Neighborhood Cleanup Program). "Help us help you," they asked. They asked whether the city should restrict the pick-up of construction materials, whether they should enforce the separation of green waste from other trash, and whether the cleanup program was worth the $27.36 a year each resident paid. (I did not know I paid for this program whether I produced an annual pile or not.) When I returned my survey, I attached a long letter with reasons to end the program and alternatives for the money. The city should replace the oversized dump piles with a solid city infrastructure of reuse and sharing, accompanied by extensive publicity. It should take seriously its role in educating citizens on the limitations of recycling and the enormous costs to the earth of a disposable culture. All citizens should be *required* to compost yard and kitchen scraps and to separate aluminum, glass, and paper for recycling. Personally, I'd ban one-time-use plastic disposables or highly tax their use.

~gggee~

The granddaddy trash day helps us forget and forgive our garbage; it doesn't require us to consider garbage long *before* we throw it out. One day it all disappears—in whatever amount we have—and the streets are swept clean and the trucks drive away. In our everyday lives, cultural perceptions of garbage as being value-less cloud the true consequences of tossing it in the street for one week, or tossing it in the bin and closing the lid the other fifty-one weeks.

With easy and cheap garbage, we will never learn to live with the earth in a cycle of birth, death, and return. Everyday goods might seem detached from the physics of embodied energy or from biological lessons of decomposition, but that detachment is as temporary and transparent as gossamer wings. In *An Ontology of Trash*, philosopher Greg Kennedy describes "indivisible unity," meaning that a person cannot move independently of the world and everything in it. Our mere "being in the world" means we are already returning to our bodies and lives that which we think we have discarded: chemical loading that begins in the womb, plastic bits that choke birds and babies, metals mined and manufactured worlds away that compromise water and habitats, and all of it lofting greenhouse gases into an atmosphere and oceans that circulate, eventually, right back to us. There is no *away* to send our garbage; it never truly leaves.

Our mounds of garbage provide priceless lessons about how we value the objects that surround us. These goods carry stories of complex psychological insecurities that send us to the store time and again, trying to differentiate and elevate ourselves with products instead of personal interaction and contemplation—well-having instead of well-being. When goods fail to mend and heal, we separate from them, deem them valueless. Tragic, because at their most elemental and material level, these products carry within treasures from all corners of the earth.

While walking the dog the night before the trucks came for the piles, I helped two kids drag a metal fencepost attached to a large concrete plug to a pile of odd-sized cement chunks and what was once a hedge. As I walked the couple of blocks home, the first crickets chirped. Suspended dust particles from the windy day softened edges in the fading light. I waved to a couple sitting on their porch; the man raised his beer bottle in reply. Each time I passed a trash pile, it seemed to sigh a bit, with ancient energies pulsing through. The press of dinosaur feet on mounds of greenery, their bodies liquefying into black carbon. The weight of ocean water on delicate sea creatures, and the mass of forests pressing on the earth. The piles sagged with life and death, and teamed with the possibility of life once again.

A Regular Day
for the Moon

I was pouring coffee when I heard that NASA had bombed the moon.

The radio reporter said that at 7:31 a.m. Eastern Time, NASA crashed a heavy piece of space junk into the moon's south pole to find out if the dark, cold craters contained ice. The two-ton empty rocket part smacked the surface at five thousand miles an hour, kicking up a billowy cloud of dust. Four minutes after impact, a second spacecraft with cameras and scientific instruments descended through the dust, checking for water and sending data and live footage back to Earth, before it crashed into the moon as well.

The reporter said the entire spectacle (broadcast live on NASA's website) wasn't all that dramatic. No flash, no bang, just gray craters that slowly got bigger and bigger, then a blank screen. This space mission was a search for water, perhaps frozen in eternal darkness in lunar craters. The mission succeeded and collected good data. "Game over," the mission manager said as I envisioned a pinball machine flashing those words.

The reporter, who acknowledged that people were upset that NASA bombed the moon, said, "The moon gets hit by meteorites all the time. It has no atmosphere, so it's constantly getting pummeled by things from space. . . So you know, it's just a regular day for the moon."

I set my coffee mug on the counter and looked at the sky above the house next door. Just what was a "regular day" for the moon? A few astronauts placed boot prints on the lunar surface, planted a flag to claim it, and pocketed some moon rocks, but surely that doesn't amount to "knowing" the moon and what a regular day there is. Slamming a rocket into the moon's backside seems a crude way to try and know something better, like kicking a creature with your boot to see what it does. The only moon most of us know is that sweet bright lantern in space, the one that Hollywood uses to evoke moods both romantic and foreboding.

Was the moon waxing now, or waning? In a big city, I lose track. Some evenings, I catch its full face rising behind the Wasatch Mountains or glimpse it sickle-thin in the morning sky. "Oh, there's the moon," I think. Our moon, 240,000 miles away, a faithful silent companion of our planet.

I do not know what a regular day on the moon is like, but truthfully, I'm not sure I know what a regular day on Earth is either. Culture demarcates weekend days from workdays and the transition between is TGIF. I seem to progress through the workday's hours according to lines filled in on my calendar: office hours, meetings, class time, mealtimes, appointments, exercise time, bedtime. Selfishly, the day feels like it revolves around me and the humans in my life. It's a day in Salt Lake City, not planet Earth. Though I will hike in the mountains surrounding the valley this weekend, they too seem like a world so separate and distinct from my regular workday world.

I rinsed my oatmeal bowl and stared at a mourning dove perched on a neighboring roof. How superficial is my notion of a "regular day"! I interact with all the systems on this planet and yet know little of how the Earth shapes my motions and the motions of everything around me. If I stepped back and viewed this day— this one regular day—with a macroscopic lens, would I feel differently about life on my planet? Would examining the taken-for-granted bits of a regular day change what I see as my home and my relationship with it? Thus began the experiment: to pay close attention to the regular day dawning before me.

I packed my briefcase and slipped into my coat. On these mornings in early October, I rose in darkness, ate breakfast in the gathering light of a flat, gray sky, and departed the house when the sky had just a hint of color and depth.

The routine of this schoolday—of every day on Earth—commences with gathering light. Earth, after spinning us away from the sun while we sleep, brings us back to face it once again. The twenty-four-hour rhythm, this primordial cycle of light and dark, is the fundamental context for life, and it governs absolutely all things on this planet—humans, plants, other animals, weather, oceans. Thus, "day" is both a unit of time and an astronomical phenomenon.

Next time I go to bed thinking I didn't really go anywhere that day, I'll remember that while standing perfectly still, I'm flying through space. The Earth rotates on its tilted axis once every twenty-four hours, so a spot on the equator travels about 25,000 miles through space every day, at a dizzying 1,000 miles per hour. When I learned that I looked at my feet, amazed at the solid sensation, held firmly to the planet's crust by gravity while I sped through another day.

Compared to that dizzying daily ride, a moon ride is as slow as a carousel, rotating at just ten miles an hour. If a "day" is how long it takes for a complete rotation on an axis, a regular day on the moon is about thirty Earth-days.

As I walked to the bus stop, I exaggerated my shuffle through leaves on the sidewalk. A scrub jay hopped branch to branch through a pyracantha shrub, plucking the orange berries. Magpies squawked from the wires. I heard the four-lane street at early rush hour long before I saw it. I've walked this quotidian walk a thousand times; I was heading to work, thinking about the university classes I'd teach, about a review deadline. Yet the foundation of all my daily patterns—waking, eating, walking, working—is the daily spinning of my planet through darkness and daylight.

Despite all the distance we Earthlings travel each day at such great speed, just one of those cycles—a day—is fleeting. Some days seem to simply evaporate. Aptly, the word *ephemeral* comes from two Greek words meaning "lasting but a day."

At the bus stop, I cradled my coffee mug. The faint gray-blue sky pulled itself from the surrounding clouds. Sun now lit the lower flanks of foothills facing southeast. Further south along the spine of the Wasatch Mountains, light intensified behind the crest, washing the sky from gray to rose to the palest of blue. Suddenly the sun cleared the ridge and forced me to look away. With this first blaze, shadows sprang forth—roof-lines across sidewalks, dappled crowns of maple trees stretching long across lanes of traffic, and the pointed tops of picket fences cutting the grass with pinking shears. The crisp morning grew golden.

My undergraduate photography professors assigned picture-taking at different times of day to demonstrate that the angled rays of morning and evening create dramatic light-dark contrasts with depth and dimension. As science writer Michael Sims said in *Apollo's Fire*, "your perception of shadows is an inescapable part of your experience of the day." When light strikes an object, it casts a shadow. Except for high noon in midsummer, we make sense of and perceive our visual world—its depth, color, distance, texture, and speed—as much through shadow as through light.

In the minutes before the bus arrived, the sun enlarged and rose dramatically. Except, it had not "risen" at all. Contrary to all the songs about sunrise and sunset, the sun does not rise above a static Earth that is waiting for it to return. Instead, the Earth revolves to face the sun once again. What looks like a sun rising over a mountain is actually the Earth rotating *down* and making the sun seem like it's going *up*. What an optical illusion, a grand contradiction between sensation and fact. (It also illustrates how we tend to think that everything revolves around us.)

Over the centuries, all creatures instilled deep in their bodies the Earth's patterns of light and dark and woke, slept, and ate on a natural schedule. As humans evolved, we developed enzymes, pigments, and hormones to automatically regulate and sync our bodies and brains (and things like blood pressure, temperature, and—of interest to my students—test-taking ability) to the ever rotating light-dark Earth. Scientists call the resulting rhythms circadian, meaning "about or around day." When something disturbs these daily rhythms (like long-distance travel), our bodies get out of whack. Even if we think we're subverting day-dark cycles with artificial lighting or caffeine, our bodies attend to the spinning Earth.

On the bus to campus, I did not read my newspaper but gazed out the window. In addition to light returning each morning, what makes a regular day regular is air; I drew in a deep breath. Air is just as essential but even more taken-for-granted because it's invisible (unless smoke or pollutants are added to it). No one knows of any other planet (or moon) that has an atmosphere with enough oxygen for us to breathe.

Earth's atmosphere is a thin layer, likened to a dollar bill wrapped around a classroom-sized globe. The gas mix is three-fourths nitrogen and one-fifth oxygen, with bits of water vapor, argon, neon, carbon dioxide, helium, and other elements. Without this protective blanket of gases to regulate how much heat from the sun reaches Earth's surface and how much escapes back to space, our planet would be inhospitable, perhaps like a too-cold Mars or a too-hot Venus. In part because Earth is the most massive of all the inner planets, it can hold more atmosphere to its surface by gravity alone. The same gravity that holds my feet firmly to the bus floor also holds air near the Earth. The atmosphere's layers change as they extend toward space. Weather occurs in the lowest layer and jets fly in the second layer where protective ozone also lies. Meteors burn up in the third layer and the space shuttle orbits in the fourth.

My regular day is actually pretty phenomenal for it is utterly unlike those on any other orbiting body, including the moon. Without returning light and an oxygen-rich atmosphere, none of what I saw out the bus window would exist. Now I understand why people worshiped the sun. Already, this day felt different.

As I worked in my office for several quiet hours before class, I watched chickadees flit through the branches of a viburnum, its leaves still deep green. Light-dark cycles affect plants, too. Charles Darwin, whose experiments with plants were some of the earliest work on light-dark rhythms, found that a plant reached its tallest point in late morning between 10 and 11 (and sank to its lowest point between 3 and 5). So the viburnum was having its most productive time of day, just like me.

A plant needs to compensate for its rooted life, so it has developed complex systems to sense, seek, and capture light. In Italy years ago, we drove by an enormous field of sunflowers early one morning, and when we passed by again at noon, the field was entirely transformed: all of the thousands of yellow heads had turned as if a captain said *about face!* Aptly named, sunflowers engage in heliotropism, or "solar tracking," and turn their buxom heads to follow their food source, the sun.

In *What a Plant Knows*, biologist Daniel Chamovitz explains that plants monitor the visible environment all the time: "Plants see if you come near them; they know when you stand over them. They even know if you're wearing a blue or a red shirt. They know if you've painted your house or if you've moved their pots from one side of the living room to the other." Plants "see" in the sense that light stimuli is received, interpreted, and in some manner, reacted to. "Eyes" in their seedling tips have phytochrome receptors, a light-activated switch that senses light and determines its color. While we see daylight as colorless, plants see light in a rainbow spectrum of differing wavelengths, using blue light to know which direction to bend and red light to measure the length of night.

❦

My shoes clicked down the sidewalk to my first class; neatly shorn grass lay on either side and a brick planter box of shrubs stood on the right. These plants were a planetary accomplishment, made possible not just by the university's grounds crew but by Earth traveling and tilting through seasons with warmth from the sun and gases swirling overhead. But why do I think of these plants differently than the shrubs and pines in the Wasatch Wilderness rising on the city's east side? Culturally, we value and treat the two as two separate kinds of "nature."

"Is there nature in this room?" I asked the students in my environmental communication class. Thoughtful silence; one student looked instinctively out the window. Then, one student held up a water bottle. Another pulled lunch from her backpack; "Well, food comes from nature."

"Keep going," I said.

A student tapped her desk, "I think this is wood."

"Oh, the air!" one exclaimed.

"How is glass made?" another asked, to which a student responded, "I know there's sand and a lot of heat goes into it."

Rapid pointing followed—steel, plastic light fixtures, electricity, jeans, laptops, shoes, cell phones—and speculation as to what bits of nature were inside them.

"Do we have any other choice, any other place to get things other than from nature?" I asked. It was hard to put our arms around the idea: absolutely everything we touched, everything we "owned," had its roots in nature.

I asked, "So why don't we think of nature as being here in this room?"

One student said, "It's just in the background—it's not why I'm here," to which another replied, "But nature is the reason this classroom is even here, so it kinda *is* why you're here."

A young woman who did not often speak said, "I guess it's because what's in here is like dead nature—the good stuff is out there, still alive."

A young man said, "It's that nature-culture divide we read about—we think that civilization and buildings and stuff are separate from nature, like they don't have anything to do with each other."

"So where do we think nature is?" I asked.

"It's where I go on my weekend," a young man grinned.

Sometimes I'm guilty of this, too, thinking that real nature is where I hike and seek serenity, that it is somehow separate in both time and space from my regular life. Scott Hess, an English and environmental studies professor, wrote, "When most people in North America today hear the word 'nature' they don't think of the dandelions in their front lawns, what they had for dinner last night, or the contents of their garbage bins." We tend to think that where we live our lives—if it is "nature" at all—is second-class nature. Real nature is where people are not; real nature is not "dead" nature.

It's curious that this hierarchy is based in part on the degree of human touch. The more we domesticate and alter nature, the less we esteem it as nature—but shouldn't the opposite be true? If nature is in my jeans (and my genes) and this laptop and my sandwich, I should feel utterly close and connected to nature for it covers my body and fills my stomach.

Living according to a hierarchy of where "real" nature is lets us believe we are somehow unbound by the physical world in our daily lives. We reckon that the "new economy" is magically separate from nature and not created entirely from it—like the sixty-six minerals from twenty-seven countries needed to make a laptop. What unites the two natures is that both were created by the spinning Earth and the daily light it brings to nature near and far, nature pristine and transformed.

⁓ℰℓℓℓℓℯ

Too often, I eat lunch at my desk. But today's macrolens on the "regular day" drew me outside. On a bench beneath a distinguished sycamore, with bark patterned like large color-ful puzzle pieces, I pulled out my sandwich and watched the wispy cirrus clouds over the Oquirrh Mountains to the west. It was just past noon, the pivot of the day when my particu-lar place on the planet slid over the cusp from a.m. (ante meri-diem, before noon) to p.m. (post meridiem, after noon). The October sun slanted under the branches and found me. I drew a long, deep breath.

The leaves above me rustled softly, as if sighing. In a sense, they were. The tree absorbs the carbon dioxide I exhale and expels oxygen—and a single, mature leafy tree like this one sup-plies enough oxygen to "feed" me and another person for a year. The bigger and healthier the tree, the more oxygen it produces.

The sycamore's toothed leaf edges were browned and curled, the sign of a typical hot, dry Utah summer. When these leaves turned color and fell, their photosynthesis would cease, but the tree would continue to "breathe." All parts of a tree respire, day and night: aboveground leaves and branches get air through their pores, and roots pull air from the soil.

I closed my lunch bag and drank deeply from my water bottle, eyes closed against the sun as I tilted back my head. Trees drink through osmosis, their roots absorbing water and nutrients and pulling them up to the highest leaves and branches. It varies by tree, but this sycamore transpires about one hundred gallons of water each day; most of that water eventually evaporates and pumps moisture into the air from its leaf pores. I lingered, staring up into the branches, imagining the tree and me breathing and drinking in this day. I walked by this tree every day on campus and never really saw it, this marvel of fluid transport and a high-rise for birds that breathes and thermoregulates utterly in tune with sun and seasons.

In his provocative essay "The Trouble with Wilderness," environmental historian William Cronon points to the danger of valuing a tree in "pristine" nature more than the tree in our own garden.

> The tree in the garden is in reality no less other, no less worthy
> of our wonder and respect, than the tree in an ancient forest . . .
> even though the tree in the forest reflects a more intricate web
> of ecological relationships . . . Both trees stand apart from us;
> both share our common world. The special power of the tree
> in the wilderness is to . . . teach us to recognize the wildness we
> did not see in the tree we planted in our own backyard.

Cronon urges us to abandon the dualism, the cultural map that sees the tree in the garden as less-than and the one in the wilderness as natural and wild; instead, we need to honor both as part of home.

How I value my lunch partner and all the gases and moisture and earth-skin we share is foundational to how I value all of it—the silver maple in my backyard as much as the mountain firs I will hike among this weekend, and as much as the two-by-fours that stand as the bones of my house and the papers I graded this morning. These trees stand steadfast, embedded in my regular life. For me and billions in the urbanized world, nearby trees and birds and the winds that lift them constitute our daily nature, our daily sacrament.

I remember this today; how can I remember and live it every ordinary day?

In *Braiding Sweetgrass*, Robin Wall Kimmerer, a Potawatomie Nation member and botany professor, explains that many Indians begin each day with an allegiance of gratitude to all members of the natural world. While the traditional school-day recitation of the Pledge of Allegiance honors fidelity to a political system, Kimmerer sees the boundaries of what she honors as far bigger than the republic—a pledge of interdependence, an allegiance to a democracy of species. When feet first touch the earth in the morning, before school, before meetings or gatherings, words of gratitude are spoken—known in the Onondaga language as the "Words That Come Before All Else." She described one recitation by children in a nearby tribal school in upstate New York. A few students at a time stood and sent greetings and thanks (in their native language) to one of many groups—other people, waters, fish, medicine herbs, birds, plants, trees, animals, four winds, moon, stars, and sun—acknowledging in turn the function of each and why they were grateful for it. After two girls recognized water, all students called in response, "Now our minds are one." A little boy came forward:

> *Standing around us we see all the Trees. The Earth has many families of Trees who each have their own instructions and uses. Some provide shelter and shade, others fruit and beauty and many useful gifts. The Maple is the leader of the trees, to recognize its gift of sugar when the People need it most. Many peoples of the world recognize a Tree as a symbol of peace and strength. With one mind we greet and thank the Tree life. Now our minds are one.*

Words That Come Before All Else: a collective agreement to be grateful for all that is given by the Earth, and the reciprocity and responsibility that come with each gift. Kimmerer said you can't listen to this thanksgiving address without feeling wealthy. It reminds you that you have everything you need, which invites an outlook of contentment and respect for all of creation. She finds it strange that the Pledge of Allegiance, which is all about love of country, makes no mention of the country itself.

I packed up my lunch bag. The Potawatomie prayer of thanksgiving dissolves the hierarchy between wild, domesticated, and dead nature, and between humans and nature. They are indivisible. We all drink in air and water and turn our faces to the sun. I am grateful to be here, for there are no trees on the moon.

✳

After my second class, I walked uphill through campus. The morning's thin cirrus had organized into towering puffy mounds, announcing a shift in weather. Wind skittered leaves, tousled my hair, and buffeted magpies who used their long tails to rudder down to the grass. Wind reveals movement, but in fact, *everything* moves, including the inanimate. Air and water that appear perfectly still nevertheless experience a continuous mixing of gases and particles—hydrogen dancing with dust, moisture mingling with methane. Deep inside all matter at the molecular level is a virtual cosmos in motion. As science writer Michael Sims noted, "those uncertain electrons whirling *somewhere* in Zen emptiness have made us realize that the very stones beneath our feet are as loosely woven as lace." The energy of dancing wind is palpable compared to all the energy whirling below our level of awareness, and it reminds me that the sum of nature is comprised of far more than its breathing and respirating members.

The constant movement and energy of the physical world means that the Earth (and also the moon) are not the passive objects that culture perceives them to be. The moon is incredibly powerful—not just on our psyches, hearts, and minds through the centuries, but in pushing and pulling Earth's tides, and moving the menstrual blood in women. Because our satellite moon is so large, some describe the Earth-moon relationship as a two-planet system. Indeed, the current theory is that the Earth and moon were born of the same force. About 4.57 billion years ago, a planet-that-is-no-more struck an Earth that was about 90 percent of its current size with the force of trillions of hydrogen bombs. Debris from the impactor planet rained to Earth, and some was boosted into orbit by gravitational torque and eventually amalgamated to form the Moon.

I now believe that Moon deserves to be capitalized. From now on, I shall.

Despite a shared material ancestry, the surfaces of these two bodies could not be more different. The Moon is poor of life; the Earth is rich with it. Meteors (and rockets) hurl to the Moon's surface because it has no atmosphere to protect it. Earth's protective atmosphere makes it a Goldilocks planet—not too hot, not too cold, a place pregnant with green growth and life.

What is mussing my hair as I walk through campus is—come to think of it—my planet. As Earth turns toward the sun, some regions heat up, and heating-cooling disparities help spawn air movements (like wind) and precipitation. Energy from the sun is the driving force for the planet's hydrologic cycle, the endless cycling of moisture from precipitation to evaporation.

On the Moon, a regular day is entirely without wind or clouds. Without an atmosphere to churn gases and moisture into magnificent mosaics of clouds, the Moon is quiet and still. Without clouds, there are no colors at sunrise or sunset. Without breeze of any kind, the footprints left by astronauts decades ago (and the space junk just crashed there) will remain intact for millions of years as if flash-frozen on the lunar surface.

I opened the door to my office and thought, where did this day go. *Tempus fugit*—time flies. My friends and I agree that time does indeed seem to fly faster as we age. Sims explained this well in *Apollo's Fire* as the sheer accumulation of day-night cycles: "It is the undeniable realization that every day we live constitutes a smaller percentage of the accrued experience with which we awaken each morning, and therefore seems proportionately a smidgen quicker and smaller than the day before." The telescoping of these cycles over a lifetime is both cautionary for how to live them, and reassuring that days will continue to spin by eons after I am gone.

Even as they fly faster, I find it remarkable—improbable even—that we have regular days on this planet at all. The impactor planet that struck Earth is called by some Theia, in Greek mythology the mother of Selene the Moon. Since their explosive births, while the Moon looked serenely on, almost unchanging, Earth went through nearly its entire history. What a happenstance that all the pieces came together—just enough carbon, hydrogen, oxygen, water—to support life on Earth but not the Moon. The regularity of our planet—sun, air, seasons, growth—provides a reliable platform on which to enjoy both the regular and the momentous events of our lives. Of course, impermanence and change are also part of every regular day, with its innumerable and complex patterns of life, death, and transformation. But at times, the constancy of a trillion new dawns and dusks seems far easier to hold onto.

After dinner, I leashed the dog and we entered the gathering dusk, walking west toward the high school and large park beyond it. Maddie paused to read scent marks on what seemed like every other tree; I used the pauses to gaze skyward. The weak windy front had departed, the clouds dissipated, and the clear sky faded to pale golden in the skewed rays of the setting sun. The Moon—a waning gibbous, I saw on my calendar—would rise at about 11.

Unlike my ancient ancestors, I don't need to retire to my shelter by dark; streetlights would soon illuminate sidewalks for our evening stroll. In just a hundred years, the incandescent bulb has shrunk darkness incredibly: streetlights, headlights, porch lights, billboards, parking lot and business lights, and, soon, holiday lights. It's as though we are trying to subvert Earth's spin into darkness and turn night into day. Our nights were lit originally in the name of safety and fear, and now also by a desire to stretch the day and remain active at night.

Darkness is important to all creatures with eyes. Bright night skies confuse hundreds of North American bird species who migrate at night and use the Moon and stars to navigate. Sometimes whole flocks crash into night-lit towers and smokestacks. Too much light also harms the nocturnal habits of sea turtles, frogs, bats, and other creatures, not to mention humans' biological clocks. Because cities like mine glow their locations so brightly into space, two-thirds of the world's population cannot see the Milky Way, that creamy, dreamy ribbon of stars that evokes as much inspiration as the Moon.

When I turned my key in the lock, it was as dark as this city ever becomes. Though we tend to think of Earth's darkness as a matter of hours, its darkness also has a size: over ninety million square miles at any one time. The people, birds, and animals in that dark expanse were slumbering. Even plants "sleep," folding their leaves and closing their pores. For humans in the dark expanse, darkness regulates their bodies: blood pressure and body temperature decrease, the heart slows, and the body repairs and rests.

In bed, my attention wandered from my book and tea to my journey through this regular day. I breathed 23,000 breaths. My heart beat roughly 115,000 times. I drank a half-gallon of water. Everything I touched at that moment—the tea mug, the air in my lungs, the sheets on the bed, my pajamas—were embodiments of everyday nature. All were made possible by the spinning Earth and the magnificent atmosphere that encircles it. I felt oddly comforted, knowing that like breathing, I do not have to think about Earth's systems: they endure, proceed, and regulate life, nature, me. But now I feel like I ought to say thank you.

I left my bedroom curtains open, hoping to see the Moon; it was like wanting to glimpse the face of a loved one to see for yourself that she's okay. Dozing and drifting, I daydreamed I was on the Moon. The surface was a dry, lifeless monochrome crust with no wind, no green cheese. The secondhand sunlight glared brighter and harsher because no atmosphere scattered the light. As I walked over a moon ridge, I saw some olivine, a lovely light green rock, just like the olivine mixed with quartz I once carried home from a Wyoming backpack trip. Mostly, I saw dark plagioclase feldspar and igneous rocks.

From the Moon, I gazed back at Earth—the Blue Marble, so named by the Apollo 17 astronauts who were honored with the first such view of our very old planet. From space, the beautiful sphere with wispy white clouds above oceans and continents looked perfect, isolated, fragile. It deserved Words That Come Before All Else. My daydreamed Moon-day was little changed from a Moon-day several billion years ago: geologic and biologic quiet. The same cannot be said for a regular day on Earth.

Of course, what is "regular" on Earth is always in flux: viruses morph, volcanoes erupt, mountains erode, species evolve and disappear, the climate cycles. But human impacts are of an entirely different caliber, as evidenced by the new space junk lying mute on the lunar surface. Recently, scientists acknowledged the scope of human influence by proposing to name the current epoch the Anthropocene; in a mere blip of geologic time, this one species has significantly altered the planet's functioning and climate for all species.

I bristle that the Anthropocene designation codifies and makes "regular" our colossal impact on the planet. The National Climate Data Center similarly codified and camouflaged the human-driven change by calculating "new normals"—thirty-year average temperatures, rainfall, and snowfall in thousands of regions throughout the country. What was considered a regular January day in, say, the Upper Midwest in the 1970s, is now two degrees warmer—but the new "normal" high temperature on the daily weather report disguises that warming.

But we do not need science and numbers to confirm that what we see and experience around us is not normal, not regular. Winters without snow (or cities hammered by it), or winters where it's balmy in January and blizzarding in March. Temperature record upon record upon record. Tornadoes in winter, fires in spring, and 100-year-floods every decade. Birds arriving a month early, gardening a changed enterprise, and ships soon traversing the Arctic. We can tell it in our bones and when our seasons do not match the months on the calendar. Instability is the new regular.

For many plants, birds, animals, insects, and fish, a small temperature bump is like living on a new planet. Very slight shifts to regular days—in temperatures and what temperature changes do to moisture patterns—have immense and cascading impacts. Warmer stream temperatures mean salmon do not spawn, which affects all creatures with whom they interact, from scavengers to insects to ocean dwellers to humans. Consider the threads connecting grizzly bears to a key calorie-laden food source, miller moths. Shifts in seasonal timing, temperature, and moisture across hundreds of miles and multiple ecosystems can break the thread—food for the caterpillar, nectar for the moth, weather the dispersing moths encounter, and temperatures on the rocky mountain slopes where they burrow and await the bears, who eat them each August by the hundreds of thousands.

I and other privileged humans can adapt. But I grieve that human hubris has changed life and its possibilities for billions of Kin who share this planet. I grieve for creatures who know their home places are no longer regular but lack options. Seabirds starving because warming ocean surface temperatures are sending fish too deep for them to catch. Pikas succumbing to new heat in their cool mountaintops. Aspen trees suffering what one scientist called heart attacks—the ground so dry that air bubbles block the tubes that transport water up the tree. And all the creatures—human and otherwise—who must flee rising ocean waters, the too-dry ground, the burning forests. What do all of us do when our regular, our quotidian, has been so altered?

This week in my environmental studies class, we were discussing water shortages and droughts in the West, and in a tongue-in-cheek way I brought up NASA's search for ice on the Moon. Near the end of a lively discussion, a young man who was usually silent in the large class raised his hand.

"But we might need that ice at some point," he said. "I mean, look what's happening down here on this planet. We might need to, you know, escape . . ." The class was silent.

After learning about a regular day on the Moon, I know I do not want to escape to there. I am an Earth citizen, a label larger and more profound than Salt Lake resident or U.S. taxpayer. I am governed fundamentally and eternally by this spinning, tilted orb that carried me 25,000 miles through space today, gave me darkness to rest, and brought me sun and wind and joy.

In an unexpected way, my macrofocus on Earth today gave me thanksgiving for the smallest and most ordinary bits of it—microbits that must be connected to the macrobits before we value and see it all as one nature. Connect the heat in my hot water to the ancient fossilized sunshine that powers it. See the apple as the product of soil created by a billion organisms and photosynthesis by a magnificent tree. See my TV as a travelogue of materials from nature worldwide. Only when we dissolve the hierarchy of where and what nature is can we reimagine how we use and treat it in our everyday lives.

Science writer Sims concluded that the real purpose of a day is to remind us of our place in the cosmos and in nature. What allows me to sleep this night of my regular day is the image of Earth turning down to greet the sun once again tomorrow. Each dawn is a new chance to know and love my home planet once again and heal my relationship with it. That is both regular and transcendental.

⊙ ⊙ ⊙

⊙ ⊙ ⊙

⊙ **4** ⊙

Out of the Woods

"He's out of the woods," the emergency room nurse said, "at least for now. We'll know more tomorrow after he's had more tests. But right now, your dad is alert and stable. He's able to respond to commands but his speech is pretty garbled."

The phone line hummed.

"We'll know more tomorrow, but he's out of the woods for now," she repeated.

"Okay, thanks," I said and hung up. Hot winds whipped the maple tree in my front yard as the edge of a front crossed the Salt Lake valley.

What the nurse (who had just come on duty) did not know was that hours earlier my dad was found in the woods behind his house where he collapsed. On a 100-degree Iowa August day, he struggled desperately to get out of the woods, pulling himself along on saplings and roots inch by inch, dragging his paralyzed right side through a tangle of branches and barbed wire, pushing off on his good left foot. He lost his glasses and one shoe. He ripped his khakis, his dress shirt. He lay there for hours—the police and paramedics said perhaps eight hours, maybe eighteen—unable to call out. When friends searched a second time, he managed to raise his left arm.

In the year since that day, my ears alert to the phrase "out of the woods." The weatherman says we're not out of the woods from the latest Pacific storm. The economist says we're not out of the woods in the housing market. My mechanic pronounced after repairing one tire that I wasn't out of the woods because the other three were wearing thin. The cultural phrase conveys that woods are full of danger and difficulties, a place one must leave to be safe and secure.

Yet for my father—and for me—woods were a sanctuary of great solace and joy. The 1960 rambler my parents built south of the Iowa State University campus was surrounded by woods, woods more familiar to us kids than any playground. Yet, there is something about woods and what they culturally represent that begets this enduring metaphor, which has been uttered through decades, perhaps centuries. Words connect us to the larger world and shape our sentiments (good/bad, wise/foolish, safe/unsafe) and our actions. Lifetimes of associating woods with dangerous places to escape must lead to some degree of consequence, such as "taming" woods worldwide for their treasures, and for our desire to refashion them into strip malls or palm oil plantations. But at the root, "out of the woods" speaks to the uncontrollable nature of nature, and of being mortal humans—chiefly, the terror of death.

⁓ℱℱℯℓℓ

A few hours before I spoke with the emergency room nurse, an Ames police officer called to say they had just loaded Dad into the ambulance.

"We don't think it was foul play," he said, "but he's in pretty bad shape, lots of blood."

The picture was unfathomable. It was Tuesday, the second day of fall semester at the University of Utah where I taught, and Dad had visited my mountain cabin in Wyoming just three weeks before: we hiked, stained the deck, and ate trout. He flew out every summer to spend time with me in the dense and peaceful woods. The policeman's call began a flurry of calls with two shifts of emergency room nurses and doctors, and with my brother Jim in California. When I collapsed into bed, I listened to the dry prickly winds, picturing the Iowa woods with basswoods and shag-bark hickories stretching north beyond the house, and my father lying amongst them.

⁓ℱℱℯℓℓ

When I was a naturalist for Olympic National Park in my twenties, each week I presented a campfire talk titled "Wilderness." I drew campers in by lighting a big fire with a flint-and-steel. Then I clicked through sublime slides of rainforests and glaciers and explained that *wilderness* came from an old English word *wild-deor,* meaning "of or about wild beasts." In between shots of bright-white mountain goats and marmots, I extolled the extraordinary wooded wilderness.

I was naive then about the ancient biases against wild woods that were deeply rooted in the psychology and history of our ancestors. When William Bradford stepped off the Mayflower, he called the woods "hideous and desolate" and began a tradition of repugnance-driven "taming." In *Wilderness and the American Mind,* historian Roderick Nash notes that except for brief movements like Romanticism, most of European and Western history considered wild nature a moral and physical wasteland, a fearful place of sinister monsters and degenerate savages: "Safety, happiness, and progress all seemed dependent on rising out of a wilderness situation. It became essential to gain control over nature."

~୨ଌୠଡ଼ୠ~

Early Wednesday morning, the neurologist called. Dad had had a cerebral vascular accident.

"A CVA is a very large stroke," Dr. Spencer said. "One-third of his brain is damaged. He cannot use the right side of his body and he can't talk. He's not following commands like he was, even earlier this morning."

"Oh my god. . ."

"He's seriously ill and you'd better come. There's an immediate danger of swelling, and with a stroke this large, that's unpredictable and it could be lethal."

"Oh Daddy. . ."

"And, in about seventy-two hours, we'll need some guidance. His health is good, but his age is working against him. Even if he survives this, I suspect his language will remain impaired."

I told him about Dad's do-not-resuscitate order and advanced care directive.

"Good planning on his part," he said. "But you need to come."

Jim and I booked plane tickets, he from Oakland and I from Salt Lake City; we would converge in Des Moines that evening. As I frenetically packed and sent emails to colleagues, I thought about the eighty-seven-year-old celebrated inorganic chemistry professor—who still worked six days a week in his lab—not being able to talk. At twenty thousand feet over the forests of the Rocky Mountains, face turned toward the window, I wept.

Driving from the airport north to Ames, Bob and Jenna stitched together a story with large holes. Bob, a chemistry colleague and longtime family friend, was the one who spotted Dad's raised arm on Tuesday about 4 p.m., the second time they had gone to the house. The house was locked, the garage door open, the car inside. Dad was at work all day Monday, but when he did not show up on Tuesday a staff scientist called the house repeatedly, and then called Bob. Dad either walked into the woods Monday after work, or Tuesday before work. The briefcase he always carried and his keys were missing. As Bob sped through the sultry blackness, the stalks of corn at the edges of the headlights waved their long, browned leaves: *hurry, hurry!*

After midnight, Jim and I broke into Dad's house in the headlights of Bob's car. A dry seething cacophony of cicadas rose from the muggy blackness of the woods. We found a spare car key and drove to the hospital. As we walked from the parking garage in the buzz of street lights, it was eerily familiar. Six years before, my older brother Scott died just minutes before we arrived from the airport. And eighteen years ago, my mom died two days before our planes would land. The double-doors sucked open at our approach, blasting us with frigid air.

I took Dad's hand and kissed his cheek. He looked at us, one to the other, searching for cells in some section of his brain that would tell him who we were.

❦

The root *will* in *wilderness* refers to self-willed or uncontrollable, thus wild animals and places not under human control. Nash says some world cultures (and some religions) make no distinction

74

between (nor have words for) places "controlled" by humans versus places uncontrolled and thus wild. Chief Standing Bear, an Oglala Sioux, said of white civilization, "We did not think of the great open plains, the beautiful rolling hills and the winding streams with tangled growth as 'wild.' Only to the white man was nature a 'wilderness' and. . . the land 'infested' with 'wild' animals and 'savage' people. . . . [For my people] there was no wilderness, since nature was not dangerous but hospitable; not forbidding but friendly."

In *The Control of Nature*, John McPhee chronicles how humans have endeavored to control (and think they control) lava flows, below-sea-level lands in southern Louisiana, and the ever sloughing, quaking, and flaming foothills of southern California. In each case, humans wanted to live in places with known dangers that humans' egos believed they could subvert and make safe. Human control of nature is futile and often unnecessary, but it remains an unceasing quest.

The intense desire for a dominant human imprint over something largely not-human has always baffled me. I enjoy being immersed in the larger terra incognita and subject to rules not of my making and understanding. Deep familiarity with patterns of daylight and dark, the language of creatures and trees creaking in the wind, the smell and feel of wood and soil transforms them from places of terror and danger to ones of solace and serenity. People comfortable with wildness find it a sheltering place to contemplate the incomprehensible, like mortality.

My first summer in Olympic National Park as a backcountry ranger terrified my mom.

"Oh honey, out there by yourself, alone, and who knows what people are out there."

"Mom, I've got a park radio." (I didn't tell her it didn't work unless in view of a repeater tower.) "And I feel safer there than I do on big city streets at night, I really do."

But I have been afraid in the woods. On a solo backpack with my dog in an unfamiliar wilderness, I got terribly lost. On a backpack with my ex and brother Jim in central Idaho, we ran out of food. Cross-country skiing in a frigid mountain forest in western

Wyoming, I got hypothermia. In Idaho on a whitewater river, our canoe swamped and I was pinned against a log.

There is something about the yin and yang of fear and exhilaration when you brush against mortality that is life affirming and invigorating. Yin makes the yang more palpable, more known, and vice-versa. When I encounter dangers in the woods, the fear stretches the boundaries. If there is nothing to risk—if it's all controlled and civilized—is there anything to gain?

In the musical *Into the Woods*, the characters refreshingly seek not to get out of the woods but charge into them, where they encounter the magical and monstrous. The woods are the stage for tests and quests and where ordinary life is left behind. "Come what may, follow the path and never stray, don't be scared, be prepared," says the refrain. The woods can be a dangerous place, they sing, but also a place to be wild and abandon your inhibitions. The baker's wife tells her husband, "you're different in the woods," more daring and more thriving. Yet, the two characters who do not make it out of the woods were those who morally strayed— the baker's wife (who flirted with another man) and the witch.

On Thursday morning, Jim and I walked into the woods behind Dad's house. The south side of the house opened to the gravel road; the other three sides were ringed by woods thick enough to conceal neighbors. The hardpan earth was furrowed and deeply cracked. We brushed by limp leaves, thin and blotched with brown spots. My shirt lay damp on my back; it was heading toward 102 degrees. The thick heat hushed the birds. There was no woodsy solace here.

Broken branches and several long scuff marks revealed where Dad struggled. We thought we knew every inch of these woods but didn't remember the slight rise that hid his body from the house and from searchers. Trimmed brush lay in several piles. Dad told me he had been pruning some basswoods to give the redbuds more sun; I pictured him in worn jeans and ratty hat happily clipping away. We traversed the woods several times but could not find his glasses or car keys.

The image of Dad's struggle in the woods haunted me. I imagined him heading into the woods to see something—a bird, a bloom—then collapsing, clawing at the baked earth, not knowing if his woods would swallow him up. He lay in the blazing sun, bloodied and bruised, dying of thirst, unable to yell, unsure of rescue or escape. Dad was exceedingly strong-willed and this so uncontrollable. What was the nature of his terror here—helplessness, incapacity, death, dying alone? It certainly was not of the woods themselves: only that no one knew he was in them, which was its own terror.

The Iowa summer had commenced in June with floods and lawless downpours, turning the woods tropical and dense; by August, a sluggish stretch of 100-degree days and empty skies parched the plants and fractured the black soil. Yet from beneath the hard crust would emerge next spring through the leaf litter hundreds of daffodils and tulips that Dad had planted, blooming jubilantly before the trees leafed out. Beyond the north woods where a cornfield once lay was a retirement community where Bob and Jenna lived. When I once mentioned the possibility of Dad moving over there, he resisted; what little time he spent away from chemistry was in the yard and woods. "I love it here," he said.

Jim and I left the sultry woods and drove with the AC on full blast to the hospital. We passed a tractor-trailer parked near the football stadium with giant black letters: HAUNTED FOREST: DON'T GO ALONE! OPENS OCTOBER 5.

If I wanted to spend time with Dad as a kid, I could follow him around the garden and woods while he worked, or bug him to read us a story. Nestled on the sofa with my brothers, rapt by his deep bass voice, we learned how the mongoose Rikki-Tikki-Tavi saved the family from cobras and how the elephant got his trunk down at the banks of the great gray-green, greasy Limpopo River, all set about with fever-trees. The tales were dark or dangerous, but magical.

To German fairy-tale authors Jacob and Wilhelm Grimm, *Wildnis* had two faces: inhospitable and threatening but also wondrous and beautiful, able to elevate the beholder and provide sanctuary. Hansel and Gretel's trek deep into the forest was scary, but behold that fanciful, edible house.

From classical mythology onward, Nash chronicles the cultures that associated woods with the monstrous and supernatural. Forests had trolls, monsters, and werewolves. *Beowulf* was the eighth-century conflict between two gigantic, blood-drinking fiends. In medieval Europe, Wild Man lived in the heart of the forest far from civilization, an ogre who devoured children and ravished maidens.

Fear of the wild traveled the ocean to the New World. Puritan minister Cotton Mather warned his parishioners in 1707 of dragons, devils, flying serpents, and witches in the primeval forests, his attempt to elicit fear and godly behavior. Alexis de Tocqueville on his 1831 visit found Americans insensible to the wonders of nature, fixed upon draining swamps, channeling rivers, "peopling solitudes, and subduing nature." Nash said the American pioneer bias against the wild was twofold. First, settlers viewed it as a formidable threat to survival (rather than the very source of their survival), which engendered hostility and a tendency to view nature through utilitarian spectacles. Second, the wild was a useful symbol and enemy to battle against in the name of nation, race, and God. Despite the occasional flowering of admiration for nature (ironically, often in cities), Nash concludes that appreciation never seriously allayed the settlers' aversion. The fear of forests, even if a result of superstition and hyperbole, continues in the minds of many today.

"Hey Dad, it's Jim."

Dad's face lightened. For several hours Thursday afternoon, his eyes examined visitors from the lab and church with a silent shine of what might have been recognition. But news from the doctors was not hopeful. His swallow reflex had not returned, which was crucial for eating. His breathing was more labored. His brain could still be swelling. Still, it felt like part of our father was not lost in the woods.

About 9 that night, the hospital called. Dad's heart was racing; they called it SVA, supraventricular arrhythmia, an abnormal electrical signal that made his heart quiver and beat wildly. We alternated all night between his room and a visitors' lounge, where we slept fitfully in boxy reclining chairs. When I walked the long hallway to check on Dad, I heard his breathing from several rooms away—weighty gasping breaths, in out, in out, in out, a continual quest for oxygen. When his speeding heart spiked into the red zone, monitors sounded and nurses came, a commotion that drowned his panting. Never have I felt so helpless and scared, like I had stumbled into an unfamiliar, dark wilderness. On a face frozen without expression, Dad nevertheless looked afraid as well.

"Daddy, it's okay," I whispered in his ear, "whatever you need to do. Don't be afraid."

In my sixth grade Episcopal confirmation class, the priest assigned the topic Death for my final paper. After I handed it in, Father Paul asked me to try again. I wrote about the holy trinity, life-everlasting, other bits of church doctrine; he was not impressed with the rewrite. Death was unimaginable to that twelve-year-old.

I once asked Dad how he reconciled his science with his Episcopal faith. "The Bible is a really good story with great lessons for life," he said, "I don't take it all literally. But I do believe in a higher power." We talked about his end-of-life care and cremation but never about death and whether he feared it. All I knew is that he feared running out of time to do chemistry.

Philosophers tell us that without death, we would be unmoored and unmotivated to get out of bed. We work to stave off death with achievements: agriculture, shelter, food preservation, and one of our greatest achievements, science. Dad's chemistry gave him purpose and his skill in it left him a legacy, two factors scholars say are related to our quest for immortality.

Carl Jung called religions "complicated systems of preparing for death." Many religions teach us that each human has a soul that will live forever in a joyful state if—and it's a *big* if—she follows the rules set forth. Thus, it would seem, the more religious a person was, the less she would fear death, knowing that spiritual immortality awaits. However, it is not that clear cut. As "terror management theory" puts it, if you accept the existence of an afterlife, it may increase your fear of death because the outcome is either really good (eternity in heaven) or really bad (eternity in hell). Your teachings might have told you God was demanding and vindictive, or lenient and forgiving, but either way your final fate was unknown and a matter of faith.

On Friday morning after a sleepless night, we drove home for fresh clothes and retrieved things we hoped would reach Dad. We filled the hospital room with Chopin etudes and Mozart concertos and placed a picture of Mom holding a koala bear near his bed.

"He's not responding," I said, tears brimming. "It's not working."

Jim put his arm around my shoulders. "We don't know that, Sis, he might hear it."

Without language or facial cues, without response from the glazed eyes that stared beyond us, was it conjecture—or faith— that his brain could process these moments. We did not know whether thoughts were present but blocked, or whether there were no thoughts at all.

I sobbed and hung on Brenda when she arrived; Brenda was my brother Scott's widow, a nurse who was in town briefly and who immediately became our medical interpreter and life-support system. Dad's community from work, church, and duplicate bridge became a food-support system, delivering meals for Jim and me. My eyes were bloodshot raw and my body bore a heavy permeating ache, but the food tasted like homemade love.

After the rough night, we limited visitors to his room but passed along their wishes. When we read Dad a card from Wei Wei, his Chinese postdoctoral student, Dad pointed to the card, looked at us, pointed again, half of his mouth in a seeming half-smile. When we brought her in, he touched her with his left arm, held her hand. Wei Wei beamed at him. "You going to get better real soon, John, real soon you be good as new and back to work."

The doctors did not agree with Wei Wei's assessment. Still no swallow function. The heart arrhythmia continued though the spikes had abated. His lungs were filling with fluid, likely the onset of pneumonia. I dabbed lip balm on his cracked lips. His mouth hung open since the stroke and the constant rush of air through his throat and mouth must have parched him terribly. I envisioned him drawing down a long drink of water, his Adam's apple moving up and down, which fascinated me as a child.

The Garden of Eden was a peaceable kingdom with abundant water and food where the beasts had no "wild" nature and got along. It was a tame and gentle garden—not a woods. When Adam took a bite of the apple (and it got stuck in his throat), the pair was expelled into a wilderness cursed with thorns and thistles— the physical and spiritual opposite of the garden. In various exiles and exoduses, droughts meant God's displeasure, and rain signaled his approval. Wild places were one of God's most powerful weapons, places to banish sinners to seek penance and purification.

A group of psychologists theorized that most of what we do and most of what we believe are motivated by the fear of death. Vitamins, face-lifts, prayer, exercise, health screenings, hair dye, Viagra—all are attempts to keep us young and vital and stave off the inevitable transition from life. Psychologists say that our worldviews—like religion, nationalism, environmentalism— exist to help us cope with our fears of mortality. When something reminds us of our inevitable demise, we cling all the more fervently to these beliefs.

Past Dad's garden at the edge of the woods and past the tire swing lay Dad's compost pile, contained on three sides by thick upright boards. Dad turned it occasionally, but mostly the fruit and vegetable scraps just slowly decomposed, settling into the earth as worms and bacteria did their thing, creating rich, black Iowa soil. We used to shake our heads at our neighbors to the south who raked all the leaves from their woods, leaving a sterile duff-less surface.

Denying the cycle of birth, death, and return does not mean it will not happen. Before he died, Scott and I talked about this. We both believed our basic elements returned to the earth, but our energy could not be destroyed and lived on. We liked the Buddhist version: no creation story, no beginning, and no end—a continual, continuous unfolding and coming into being. Like compost in the woods.

༄

The fervent heat continued, making each journey from house to car to hospital a set of dramatic thermal transitions that signaled my body's travel through very different worlds. I began carrying a sweater everywhere I went.

By Saturday, hope for Dad was crumbling. His lungs had full-blown pneumonia. His heart rhythms were still wacky. His agitation worsened and he tried to get out of bed, which sounded the bed alarm. We decided to welcome any and all visitors. People wedged in the room, telling John-tales, laughing at his wit and stubbornness, his determination and courage, his Scotch frugality, his intelligence. Hugs and arms entwined me. Our eyes abided on the man in the bed who stared into the distance and breathed with all his might.

In a moment of solitude, I said, "Oh Jim, our options feel very different today."

After each test and lab report, after each conversation with his neurologist and doctor, Jim and I returned to our metaphorical decision-table, which now had fewer things left on it to consider, fewer measures of hope. We no longer spoke of whether he would be content using his left arm to spell out words on a letter board or whether he would enjoy the social environment of a nursing home. We did not believe this gustatorial fanatic would enjoy tubed puree that bypassed taste buds, even if he hadn't eaten for five days. His body was making decisions for us.

༄

Last spring I found a dead warbler in my backyard—a dark olive back with a yellow belly and gray head, a female MacGillivray's.

It was gorgeous, even in death, and it touched me that I stood in the spot where it took its last breath. I stroked its feathers, knowing I would never hold one closely again, and placed it in a crook of the juniper.

Buddhism, Hinduism, Shintoism—all profess a human-nature relationship marked by respect, where humans are a part of all nature and have compassion for all living things. Many of the world's religions grant only humans spiritual immortality because only they possess souls. An exception was St. Francis of Assisi, who believed all creatures had souls and who preached to them as equals; his posture of respect and compassion lay closer to Eastern religious traditions than his Judeo-Christian one. At the time, the Catholic Church was not ready to concede that humans were a part of nature rather than above it, and it labeled St. Francis a heretic. As Nash concluded, "Christianity had too much at stake in the notion that God set man apart from and gave him dominance over the rest of nature (Genesis 1:28) to surrender it easily." Thus, the recent Encyclical by Pope Francis is truly revolutionary; he embraced his namesake's belief that all creatures have souls and will go to heaven.

I like to imagine that Father Paul's assignment was to go into the woods like Chinese religious teachers instructed, for in the woods one could sense more clearly the unity and rhythm of the universe. In the woods, I envision the soul of a warbler singing to the great beyond and the soul of a tree as it cradles the dying bird and shades me. Warbler, tree, and woman sharing air and earth, united by our mortality. There is no into or out of the woods that I need to go.

◦≈≈≈≈◦

Early Sunday morning, a cold front broke the triple-digit heat with light rain and distant thunder. The humid haze dissipated and the distant cornfields looked closer. There was room in the air to breathe.

Jim and I wanted to take Dad to his home in the woods to die, but the following day was Labor Day, and we couldn't get a hospice nurse or a hospital bed delivered until Tuesday, which might be too late. He nevertheless needed a more peaceful place. Late Sunday afternoon, I rode in the ambulance with Dad to a hospice center in west Ames that Jim and I had visited that morning. In his spacious room, I watched the nurse remove the old dressings on his abrasions and wounds and wash lightly around them.

"You really went through a lot, didn't you John," she said softly.

She called me to the bed to point out changes in his skin color that had already begun.

"It won't be that long now," she said.

My weeping came in waves now, and over the past six days I had grown less self-conscious about it. The nurse touched my arm as she passed me.

When she returned, I said, "You're amazing. How can you do this job and not get drawn into all the grief around you? I don't see how you can do this."

She finished taping a large bandage, then looked at me. "I used to be a neo-natal nurse. When I left that and became a hospice nurse, I found that it's really just the other end of the same process. One, you usher life into this world, and the other, you help usher people out. It's surprising how much the two thresholds have in common, really. They are both such powerful transitions."

After Jim and I ate a meal delivered by family friends to the lobby, I climbed a grassy hillside behind the center and sat on a stone bench beneath a spreading oak. I called my best friend Sara and bawled in choked sobs as the sun sank beyond the golden fields of corn.

As darkness deepened, Dad's breathing that had been so heavy and labored and hard for so many days, slowed and shallowed. We put his country-western CD on the boom box. I held his hand, talked to him, cried on him, told him it was okay to go. Shortly after midnight, the last vapors left his lungs into the cool currents of night.

I will always wonder about his last moments. If the stories of the dying are true, perhaps his "thoughts" moved from terror to solace. I like to imagine that he saw fluttering hickory branches bedecked in crimson cardinals, and that light beckoned through his woods to come home.

◦◦◦◦◦

The week after his funeral, I drove in a blur to my Wyoming cabin. As soon as I stepped into the woods, a flood began that I was incapable of stopping. I cried for three hours, wandering through the trees, at times collapsing in the duff or leaning against a rough trunk. This was the last place I saw Dad, and he loved these woods, too.

In October, Jim and I returned to Ames to sort through Dad's belongings before an estate sale and to spruce up the house to sell. The woods had healed from their August scorching and the lawn was lush, deep green. Cool mornings warmed to Indian summer days, and we cranked open windows to fill the stale house with scents of leaf and earth. Maples and hickories towered over the house in crimson and gold fragrant plumes. Squirrels scampered burying hickory nuts for the season to come.

The selection of things most important to Jim and me was fairly painless and without disagreement, for which we were grateful. The most exhausting part was sorting through paperwork that plumbed the depths of our family's lives: love letters my parents exchanged, every card I ever sent Mom, records (financial, medical, sentimental), a hundred years of photos of relatives, including many we never knew we had. So very much paper. I apologized to the trees but thanked them, too.

There was a football game at ISU that weekend and Jim bought some press-on, under-eye "football blacks" which we both affixed. As I closed and locked the U-Haul door, I told Jim the eye stickers would help me be tough like a football player. We both started to cry.

Then, Jim and I walked into the woods. We opened the heavy plastic bag of ashes, cradled it, and tipped it; the gray grit spilled noisily over the leaves, marking the woods with our father. A gray streak stained where he had fallen, his carbon meeting the carbon of the earth. Another wash of gray crossed a bed of daffodils, bone meal for the blooms. We sprinkled ash behind his log pile and near the lilacs for rains and worms to blend into the soil. We sobbed. Dad and his woods were inseparable now; he would never be out of the woods again.

⊙ ⊙ ⊙

⊙ ⊙ ⊙

⊙ 5 ⊙

The Big Hum

Some mornings when I wake and the window is cracked ever so slightly and the breeze is from a certain direction, it finds me—constant and unceasing, a single medium-pitch, a faint roar: the interstate. My house is several miles from that interstate—which winds in from the salt flats, bisects Salt Lake City, then climbs into the mountains and windy plains of Wyoming—but there it is, the interstate right in my bedroom.

Of all the kinds of pollution, noise pollution is perhaps the most culturally tolerated kind. I can call the cops about a specific late-night party, but a complaint about the interstate in my bedroom would fall on deaf ears. People say they "get used to" noise and endure it as an inevitable part of city life—the price we pay to live near jobs, restaurants, and entertainment, and for which we do not consider the true costs. For me, noise is the least acceptable part of my everyday city life; it pollutes the air and shrinks the space around me. I can close my eyes to block the visual, but my ears are on duty 24/7, hearing without deliberately listening even when I'm asleep. That's tiring.

When I began investigating this topic, I had no idea whether my noise aversion was a personal sensitivity or whether it was harmful. I was certain, however, that all creatures need to be heard, to be able to capture a bit of the bandwidth to communicate and to be successful. Noise interferes with that, which must be just as true for a robin and a whale as it is for me. I learned that noise has real physical impacts on creatures' health and well-being, and that our culture accepts noise for some peculiar reasons.

In simple terms, noise is unwanted sound, sound that's too loud or won't go away. In *The Great Animal Orchestra,* musician and nature recordist Bernie Krause puts a finer point on the definition. A signal is sound that contains useful information: a crying baby, a bird's alarm call, laughter, a question, an elk bugle. Noise, on the other hand, is sound made up of unrelated sound fragments that are *not* useful—audible trash, if you will. And the source of most noise is what he calls *anthrophony,* or human-generated sound.

Some types of anthrophony are local and subdued, like physiological sounds (my stomach growling or people talking) and incidental sounds (footsteps or clothes rustling). Controlled sound (such as live or recorded music or theater) can be desired or unwanted, depending on its location and volume. But the predominant and loudest type of anthrophony is electromechanical sound: lawnmowers, airplanes, pile-drivers, the humming refrigerator, cars and trucks, a printer, the TV. . . .

Noise does not respect boundaries. And the less control you have over sounds you do not want—barking dogs, loud music, the interstate—the more it affects you, and not just your mind, but your body. Anything with ears experiences, reacts to, and can be harmed by noise—indeed, woman, child, falcon, or frog.

In a large survey of U.S. cities, the level of urban noise increased by about 15 percent in just ten years, and over a third of respondents complained about noise (generally, single noisy incidents). The largest component of the city's big hum is transit noise. But it is not just the sound of traffic received by our auditory system that is the concern, but the decidedly nonauditory reaction of our bodies to it, especially when the noise is continuous and will not go away—what researchers call chronic noise. And we can blame evolution for our involuntary reaction.

Animals (including humans) are hardwired to react to sounds and have evolved in ways that demonstrate the importance of the alerting function of hearing. Auditory organs evolved before the ability to produce sounds intentionally; that is, ears could hear before throats, vocal chords, and mouths could vocalize. And, species (including humans) commonly hear and distinguish a much broader range of sounds and pitches than they are capable of producing. The cells that move this sound information into the brain are the fastest in our bodies. Because animal brains continue to process everything that enters the ears even when sleeping or hibernating, our ears essentially never turn off.

If I'm not in deep sleep, my refrigerator kicking on in the next room can wake me up. Virtually all noise studies agree that disturbed sleep is a common result, which contributes to irritation, fatigue, and impaired performance the next day. But noise is most harmful to our bodies because it evokes a stress-response. When a sound alerts or startles an animal (or human), heart rate and breathing accelerate and the body produces stress hormones such as adrenaline and cortisol. The body is in a state of heightened readiness, the classic fight-flight-or-freeze response. It does not take really loud noise to cause these effects either. A large synthesis of studies concluded that nighttime noise exposure of only forty decibels causes direct biological effects—that is, only as loud as a humming refrigerator. And regardless of whether it's loud or quiet, noise is most harmful when your body stays on alert because the sound doesn't go away.

A friend of mine lived near the L in Chicago while in college and said she eventually got used to the noise. Another friend told me she has grown accustomed to the six-lane street outside her Salt Lake house. Some people may be able to "get used to" and become habituated to noise, meaning they change their responses like learning to sleep through it or being less annoyed by it. But chances are my two friends never acclimatized to their nearby noise, which is very different. Even when they slept, their bodies were alerted and reacted, quietly ramping up the beat of their hearts and sending stress hormones in constant circulation through their blood.

Exposure to chronic noise—particularly transit noise—is harmful. Traffic noise is generated both by engines and by tires rolling against pavement, and rolling noise is actually the louder of the two. If you think of a whisper as 20-30 decibels and normal conversation as 50-60, city traffic is 70-90 decibels, the threshold where sustained exposure can cause hearing loss. (Because decibels are logarithmic measurements, 40 decibels is not simply twice as loud as 20 decibels. For example, 60-decibel loud human speech is one million times more intense than the barely detectable 1 decibel of a leaf falling.)

Chronic traffic noise is a well-established risk factor for developing high blood pressure. Two large summaries of individual studies found that traffic noise above 60 decibels was associated with a 10 to 50 percent increase in risk for high blood pressure, ischemic heart disease, and heart attacks. That warrants a call to the cops (if not my doctor) about the interstate in my bedroom.

The World Health Organization calculated that each year Europeans lose at least one million healthy life-years due to disability or disease caused by traffic noise. That makes it the second biggest environmental health problem after air pollution. Another report from Europe found that noise from rail and road transport was linked to 50,000 fatal heart attacks every year and 200,000 cases of cardiovascular disease.

The recent flap over hybrid cars being a danger to pedestrians because they are "too quiet" needs to be reframed: The far deadlier vehicles are those making the noise.

~

We often attribute today's short-attention spans to electronics, both cell phones and screens. But a powerful underlying contributor to inattention is noise.

One warm fall afternoon, I was lured to the backyard to grade papers. I was partway through the first paper when I heard a car with a deep, thumping boom-bah-boom-boom stereo drive by. That brought the Chihuahuas and Weimaraner next door charging through the doggie door, yapping and barking. The German Shepard a block away responded. A truck somewhere started backing up with loud, spaced beeps. In the distance I heard a muffled brassy voice through a loudspeaker, like an outside intercom at a car lot or lumberyard though I could think of nothing like that close by. Somewhere to the north, a siren wailed, soon followed by a new chorus of howling dogs in all directions. A car door slammed and sparrows in the chokecherry scattered with a fluttering whoosh. People exiting the car were talking. Overhead puttered a small plane. Leaves clattered down to the patio. The neighbor with the yapping dogs turned on the stereo in his backyard. I retreated inside.

Not surprisingly, noise harms the ability to pay attention and learn. For schoolchildren, trucks rumbling by a classroom, aircraft overhead, and the noisy open design of schools are enough to raise kids' blood pressure, create learning problems, and cause "cognitive failure."

Before the airport in Munich, Germany, was moved, researchers studied kids living nearby. They found that the greater the noise exposure of children, the greater their deficits in long-term memory and reading comprehension. After the airport closed, the deficits disappeared and long-term memory improved by 25 percent. And kids near the new relocated airport? They soon developed the exact same deficits. Other studies have found a one- to two-month delay in reading age and greater psychological distress and hyperactivity for children living near airports.

When my friend Danielle learned of my research into noise, she asked if I'd come across studies of "white noise" because she had put such a toy in her baby's crib. She had good reason to be concerned: manufactured white noise is not as soothing or innocuous as I thought.

One summer on a family vacation to the Oregon coast when my brother Jim was seven, he awoke very upset. He cried that someone's air conditioner was way too loud and he couldn't sleep. It's the ocean, Dad told him. This story is all the more interesting because as an adult, Jim always sleeps with a fan on. He says he finds it soothing. He has lived his adult life in Boston and San Francisco, so this white noise seems to cover up the city noises.

White noise "works" because it masks other sounds. In addition to amplitude (how soft or loud), each sound has a frequency (low-pitched like a foghorn or high like a siren) and duration (how long it lasts). White noise is continuous noise that is equalized across the frequency spectrum, meaning that no individual high or low tones are audible; it's just one big hum. (The ocean is not white noise, for it has discernable sounds like the waves crashing onto the beach.)

Several studies of white-noise installations in workplaces found that (despite the advertising claims) the noise did not make workers calmer and more relaxed, but instead more tired and less able to concentrate and perform effectively. The studies on babies and young adults were mixed (one said it helped sleep, one said it did not, and another said white noise raised resting heart rates). The fact remains that since our ears listen during sleep to alert us to potential danger, continuous white noise means our ears are essentially forced to listen harder (which translates to increased heart rate and respiration) than they would to no noise at all.

One researcher did an experiment with rats to measure their internal reactions to white noise. She played white noise for fifteen minutes each day for three weeks to one group of rats and compared them to rats who heard no such noise. The lab's background noise level was 50 decibels (normal conversation), and the white noise was 90 decibels (like very loud traffic noise).

Although I don't know what rats consider loud, their nervous system functions similarly to ours. The heart rates and blood pressure of the exposed rats rose and remained elevated for more than ten minutes after the noise ceased. The white noise also affected another part of their nervous system, not the part keyed to fight-or-flight, but the part that helps the body "rest and digest." The small intestines of the noise-exposed rats were inflamed and swollen and had structural damage in the cells. The pictures showed big holes where some cells had separated from each other and the membrane walls.

In a follow-up experiment, she exposed another set of rats to noise for three weeks, then moved them to a quiet room for three weeks. The damage to the small intestines was not reversed by the quiet time. A stress response is necessary for survival, she said, but stress always disturbs the body's homeostasis and imposes a cost, especially when a stressor like noise is repeated or chronic.

The singular sound of white noise shrinks space—for animals (including humans) and birds—because it reduces the distance at which we can hear things. That's a problem for creatures who use specific sounds to locate each other, to locate food, or to avoid becoming food for another. If you play traffic noise to tree frogs, females cannot locate the calling males, and some frogs simply do not call as much when it's noisy. Bats who use the sounds made by their prey to catch them avoid hunting in noise; the worst type of noise for bats is white noise.

Noise also shrinks distance in a place that seems to have plenty of space: the world's vast oceans. Hearing is far more important than seeing for ocean creatures. Light doesn't travel very far or deep in water, so animals' auditory capabilities evolved to supplement or replace vision as their primary way to communicate, sense their surroundings, and find food. And because seawater is "stiff" compared to air, sound travels through water faster and farther, zipping through the salty brine at almost a mile a second, about five times faster than it travels through air. Some loud sounds in the world's seas travel thousands of miles, literally across oceans.

The communication advantages of saltwater have led humans to introduce sounds into oceans—for communication (like sub-to-sub), hunting (fish-finding sonar), and navigation (depth finders). As on land, human noise in the ocean has increased dramatically in amount, volume, and range, impacting the creatures who call those waters home. Ships towing seismic arrays blast loud sounds downward to probe under the seabed for oil and gas. These air-gun signals—over 200 decibels—last about 40 milliseconds and are repeated every 10 to 20 milliseconds for hours or days at a time. Their pulses split through the dark waters and transform those "empty" waters into places far, far noisier than cities.

Sonar works a bit like bat echolocation; sound signals bounce off distant objects and back to the source, revealing objects' location, distance, and speed. Mid-frequency sonar naval operations have been conclusively linked to changes in the feeding of blue and fin whales, in the migration of gray whales, and the reproductive behavior of humpback whales. After the U.S. Navy's low-frequency sonar was fingered as contributing to beaked whale beachings, the navy agreed to an academic review of marine mammals' response to sonar use. The review team found that the sound emitted was over 235 decibels and retained an intensity of 140 decibels over 300 miles away. Whale biologist Ken Balcomb wrote an open letter to the navy's program manager that described how low or midrange sonar caused hemorrhaging and stranding of Cuvier's beaked whales: "Envision a football squeezed to the size of a ping-pong ball by air pressure alone...[and] compressing and decompressing hundreds of times per second...in your head."

One reason I enjoy birds so much is that they are present no matter where I go, including in the city. But when the interstate noise is loud, I do not hear as many birds, and I have often wondered how noise affects birds both in the city and farther afield.

The chokecherry tree in my backyard is full of birds. Occasionally I hear a black-capped chickadee, see the dark red flash of a house finch, or in spring glimpse a yellow warbler searching my apple tree for insects, but mostly I see sparrows.

Cities tend to homogenize bird and animal communities, but I assumed the limiting factor was habitat. But for urban birds, another key factor is noise. Bird song is vital for communicating with other birds, both competitors and mates. The lower the pitch of a bird's song, the more likely the song overlaps and is masked by the city's big hum. The choices for birds affected by noise are similar to those for humans: flee the noise for the countryside, or stay and try to adapt.

One Sunday morning—Easter morning—I opened a favorite webcam trained on a peregrine falcon nest on a high building ledge in Rochester, New York. I checked the volume control because the city—usually a chorus of trains, screeching brakes, sirens, and heavy pounding—was quiet, more quiet than I had ever heard. The female was on the nest. She looked alert, as always, but I watched for something different, something calmer; I couldn't tell. I wondered if she were more stressed with the unusual quiet, like waiting in the calm eye of a hurricane for calamity to arrive.

When the British raptor expert Ian Newton spoke in Salt Lake, I asked him about urban peregrines and noise. "They get used to it," he said, "because it's constant. If they were some place quiet and then heard something loud, then they'd get alarmed."

It's what we all want to believe, both about other creatures and ourselves: we just get used to noise; it doesn't bother us. I doubted his response, knowing that even if the peregrines adjust to it, those noise levels would be extremely hard to acclimatize to. Perhaps that pigeon-filled mecca was too good of a smorgasbord to pass up, like a human who buys a house next to a freeway because the price is right; the cost of noise to person and peregrine is a revved-up internal response.

A male robin marks his territory with his song, a message that can be rendered mute by the roar of a city. One study found that in areas that were noisy during the day, robins actually sang at night. It seems smart in a way: if your song faces too much noise competition during the day, then just sing at night when it is a bit quieter. However, the night-calling robin loses rest (a big metabolic cost) and has told night predators like cats of his location.

When I read this study, I cried. Heralds of spring mornings forced to sing under the cover of darkness. A different type of Silent Spring, one poisoned by noise.

Many people have witnessed a robin hunt worms. It runs a few quick steps, stops and cocks its head for a few seconds, and then lunges, drives its bill down into the soil and tug, tug, pulls out a wriggling worm. Like all birds, robins use a lot of cues to find food: sight, smell, hearing, and vibration. If the worm is entirely buried in the soil, hearing is vital.

Several Canadian researchers did some creative experiments to test whether robins—in the absence of sight, smell, and vibration cues—would be able to find buried mealworms (smaller and less wiggly than earthworms) just with hearing. Amazingly, they could. But white noise—at only 60 decibels—significantly reduced their foraging success.

Some urban-dwelling birds change up their age-old songs, just to be heard. House finches and song sparrows have a certain amount of verbal "plasticity" and can vary their songs and sing them at higher frequencies. Song sparrow males in Portland singing in noisier locations had higher-frequency low notes. When the city songs of great tits (a common bird across Europe) were compared to great tit songs in nearby forests, the city songs were shorter, faster, and higher.

This vocal escalation has a name—the Lombard Effect—and it happens in humans and a variety of creatures, including cats, chickens, various primates, and assorted birds. Even beluga whales—those small, soft-white whales of the far north—increase the intensity of their vocalizations to compete with shipping noise. The most amazing thing to me is that this effect is involuntary; neither human nor nightingale can help but increase the intensity of their voices amidst loud noise just to be heard. It's so involuntary that Lombard Speech cannot be controlled even by instructing a person to speak as she would in silence; the voice still gets higher in pitch and volume, vowels are drawn out, content words are prolonged, lung capacity increases, and facial movements get larger.

I have spoken exactly this way in a loud restaurant. When you listen to your dining companions amidst the din, your brain works harder to combine bits of things heard with visual cues. It is very tiring to separate audible trash from significant signals. Although it's easy to add sound-dampening elements to restaurants (acoustic tile on ceilings and walls, draped fabrics, different flooring), the cultural message of a noisy restaurant seems to be "successful" and "happening." I assiduously avoid the loud ones.

~

Animals are affected by human noise even when they are far from the places humans live. In addition to needing sounds for mating and territory, creatures require interference-free communication for signaling alarm, capturing food, group defense, play, and social contact. Researchers conclude that sound is a principal way animals show emotion: grief, joy, love.

Electromechanical anthrophony changes the way creatures communicate and interact with each other in otherwise quiet landscapes, and changes where they feel safe and where they can raise young. Desert songbirds won't nest within one hundred yards of ATV trails. Aircraft overflights changed peregrine falcon parental behavior in Alaska. Prairie falcons in Idaho were flushed from nests and perches by blasting noise from dam reconstruction, and they never habituated to it; over three years, nesting success declined and some pairs never returned.

Aircraft, snowmobiles, and ATVs disturb a range of animals from harlequin ducks to mountain goats. Stress enzyme levels in both elk and wolves rose in direct proportion to the amount of snowmobile noise; enzyme levels returned to normal when snowmobiles were absent. In noisy places, pronghorn and deer spend far more time "head up" being vigilant than "head down" foraging. The mule deer population near a big oil and gas field about an hour from my cabin declined by almost two-thirds in just eight years after drilling began. Noise from drilling and traffic is shrinking habitat far more than actual surface disturbance.

For ornithologists from New Mexico to Canada, oil and gas development provides a "natural" experiment to isolate the effects of noise on bird distribution and nesting success. Noisy compressors pump oil and gas 365 days a year, while drilled-out well pads are quiet but have a similarly disturbed surface habitat.

In one study in the Southwest, house finches and black-chinned hummingbirds nested near the noisy compressors. (One reason for their tolerance may be that their higher frequency songs were not as masked.) Quiet sites had a far greater diversity of nesting birds, and nests were more successful. Noise indirectly affected pinyon pine; scrub jays (who preferred quiet sites) help disperse pinyon seeds, so where it was noisy, seeds were not getting scattered. In Alberta, Canada, noisy energy facilities reduced overall bird density by one-third, and birds' nests near the noise were less successful. The habitat lost (due to physical and sound disturbance) around the 5,000 compressor stations in Alberta's boreal forest meant a loss of 85,000 birds.

~ひなな~

Noise even disturbs the quietest places in America. For years, Gordon Hempton has been on a mission to record the world's natural sounds before they are drowned out by human noise, a project he calls "One Square Inch of Silence." His criteria: a place where you can listen for fifteen minutes in daylight hours and not hear a human-created sound. He concluded that no such places remain east of the Mississippi River and that maybe a dozen exist in the West. One is in the Hoh Rainforest in Olympic National Park, where I once worked as a naturalist.

The prospects for those few silent inches are not good. In the continental United States, 83 percent of the land area is now within two-thirds of a mile from a road. Even in large forested regions, two-thirds of the land is within a football-field-length of a road or a forest edge. Since 1970, the U.S. population increased by about one-third, yet traffic on U.S. roads nearly tripled (as did aircraft traffic and noise in the world's oceans). If traffic has tripled in the last forty years, how close will the nearest road be forty years hence?

~ggggg

With all the evidence that noise is collectively making us sick and less successful, why is human culture so big on being so loud? Noise is the second most deadly source of pollution, yet it gets little airtime.

In large part, electromechanical noise comes from sources that we associate with cultural progress, such as consumer goods, whose production, delivery, and use involves noise. And perhaps as a result of being accustomed to noise, we have become unaccustomed to (or even afraid of) silence and solitude.

For several years I gave my students an assignment: go to a spot away from humans and human signs for at least one hour and contemplate your relationship with and beliefs about the natural world. One young woman climbed a hillside outside her small town, but as she sat there, she freaked out. "It was so darn quiet," she wrote, "it felt like my heart was going to pound right out of my chest. It was just real scary being out there and I know you told us not to bring it, but I got out my cell phone and called my dad." I eventually dropped the assignment; too often, their papers dwelled on how bored and unnerved they were out there—alone and with all that quiet. I wish they could have experienced the potent contrast between city-hum and the acoustic finery of empty places.

When my dog and I walk around Liberty Park, we stop at the nation's oldest aviary to peer over the fence at the sandhill cranes. I watch them turn their long necks and scan left and right, watch them walk like ballerinas through the tall grasses near their little pond. I wonder what the big hum is like for them, if it drives them mad, how it affects their being in this place. They did not choose to be city birds. One fall day we had walked several blocks past the aviary, when I heard high in the sky the warble of four or five cranes winging south. Then I heard the call of the two grounded cranes—loud, frantic, cackling—carried across six lanes of traffic and up into the sky. It tore at my heart. I feel the desire just as strongly, to take flight from the din, and I do so every spring.

When I migrate north to my cabin in mid-May at semester's end, at first I find the silence deafening. It's as though my ears are trying to tune in and gather sounds, and finding nothing, deliver instead a constant roar of static. Soon, sounds pop through: a nuthatch flitting in the aspens, a squirrel traversing a downed log, its nails scratching lightly on the bark. The wind lifting the lower boughs of a Douglas fir. A raven calling from the sky, filling the spaces between trees.

This acoustic finery is what nature recordist Bernie Krause calls the *biophony* (the sounds of living organisms) and the *geophony* (natural sounds from nonbiological sources like wind, thunder, earth movement, and water). The sounds of rain on earth and wind on greenery were some of the first sounds on Earth's surface once the atmosphere formed, and it was around them and between these sounds that animal and human voices evolved and found their places. All organisms accommodate and adapt to the local soundscape of biophony and geophony, finding their own "bandwidth" and moment to vocalize. The intersection of animals and an alive-sounding earth is what Krause calls *spiritus* ("breath" in Latin); many cultures recognize the breath of wind as a root of spirituality, whether it moves trees in a forest or manifests in a creature's breathing.

The silence around my cabin is immediately calming, opening. It's as though the air has more space, the space more time, the time more minutes. I not only hear more, I see more, breathe more, pay more attention. I don't need to measure them; my heartbeat and respiration have slowed. Soon, my sound barometer has recalibrated so that a single, distant electromechanical sound is annoying: when the air is still in early morning, I hear the trucks climbing the highway grade three miles north.

When I'm back on campus, I am surrounded by students plugged into anthrophony from their phones and earbuds. Last week I sat next to a young woman on the bus whose music leaked from her earphones in a tinny din while she lay propped on her backpack, asleep. Many young people hiking or biking mountain trails are similarly plugged in, cut off from the rich biophony and geophony.

On evening walks with the dog in my city neighborhood, I see screens flickering in empty rooms. I hear through open windows all manner of loud companions—radio, music, TV, the blips and bleeps of electronic games. A friend's family was like that; noise emanated from virtually every room regardless of audience. "It's crazy," he told me. "It's like they gotta fill up the dead air, like it's scary or lonely or something to just be in the quiet." He was a musician to whom the silent spaces between sounds were sacred.

Acoustic ecologist R. Murray Schafer said humans like to make noise to remind themselves that they are not alone, and to remind others that they exist. Psychologically, all of us want—need—to be heard and seen. We need to believe that somewhere in all the din we can vibrate vocal chords and push air over them, and someone will not just hear but listen. We each need time to hold the listening stick. For some, revving the hot rod, yelling, blasting the boom box might seem the only way to get noticed and feel powerful.

Using noise to feel less alone is perhaps understandable, but sad. Culturally, many of us are still "bowling alone." Instead of turning on the TV for company, connect with the sounds that already surround you—the dawn chorus of birds, grasshoppers clicking in late summer, and the breeze that blows spirit into us all. It's a soundscape that humans once knew intimately and could read, each layer and cadence and resonance. As Kathleen Dean Moore and Michael Nelson remind us in *Moral Ground*, the earth "resounds with wisdom we can hear if we try, wisdom born of the longest reaches of time and space, red leaves falling from oak trees, soft rain, children coming home at dusk."

~°ᵔᵔᵔ~

Most sounds made by birds, mammals, and insects have specific communication intention, and when accomplished, the sound stops. The baby bird is fed. The pikas are alerted. The thundering wild horses have fled the danger. The dolphins have located each other. The cricket has attracted a mate. The same could be said for human-to-human communication, but the planet's noise comes not from our voices, but from our machines. And our culture is allegiant to the goods that noise creates.

Our culture associates noise with power and progress. Former interior secretary James Watt (who sought to close the EPA's Office of Noise Control in the 1980s) thought that the more noise we made as a country, the more powerful we appeared. Physically, "noise" is wasted power for it represents wasted sound (that delivers little useful information), and wasted energy (because electromechanical generation emits heat). But psychologically, we perceive noise as proportional to power and therefore enviable. The bigger and noisier the Harley, the better.

Noise-as-power is a war metaphor whose constant presence is a harmful weapon. Noise as weapon is also utterly undemocratic because the amount of exposure is biased by social class and species. Who lives in a gated community versus next to a jam-packed highway? Which species' habitats are squeezed but lack alternative places to fly or walk?

~°ᵔᵔᵔ~

Is there any movement afoot, any clarion cries for quiet, for peaceful days so robins can sleep at night? Here and there. Many cities have sought legal redress, such as noise ordinances for mufflers, truck air brakes, barking dogs, and loud noise from parties; these nuisance laws target specific loud and annoying sounds, but not the chronic components of the big hum. Other actions include "soundscape" management plans for national parks, and tightening noise limits in the European Union for cars, buses, and lorries to reduce traffic noise from 74 decibels to 70.

Bob Chanaud at Noise Free America told me that great strides were made in community noise reduction after the passage of the Noise Control Act of 1970, but since the EPA Office of Noise Control's funding was phased out in 1992, "things have decayed." Funding has disappeared and states have whittled away at gains by not enforcing or persecuting noise violations. He also cited "the power of money" in state laws that exempt shooting ranges, farms, and motorcycles from noise complaints. Chanaud's focus when he works with specific communities is largely on specific loud noises, not the cumulative impact of the big hum.

The original EPA noise office was established under the auspices of the Clean Air Act. I like thinking of noise that way, as something that fouls the air and that we have a responsibility to clean up. Beyond legal solutions, a change in cultural perceptions would help us recognize noise not as a signifier of power, progress, and companionship, but as a form of air pollution that is deadly and undermines the health and success of all residents, meaning all species with ears. But even if we change the cultural framework and embrace quiet hybrids, follow EU traffic softening, and enforce ordinances, that would fall short of tackling the voluminous sources of noise.

We create the big hum of industrial consumerism not just with transit noise, but with an ever increasing number of noisy devices and appliances: leaf blowers and cappuccino machines, hand vacuum and weed trimmers, snowmobiles and dirt bikes. Our machines provide power, comfort, convenience, reduced physical demands, fun. Our machines are powered (directly or indirectly) by the same fossil fuels whose combustion is changing and warming our planet.

As I look around my house, I realize that even if an item is not capable of making noise now, noise was discharged during every step of its journey to me. *Every* consumer product comes with noise from its production and delivery. My camera has the hiss and clamor of plastic manufacturing, the din and heat of glass grinding, the roar of looms weaving its strap. My bread contains the commotion of combines coursing the grain fields, of trucks ferrying grain to factory and bread to grocery. If my clothes, my books, my chairs—and my electricity and natural gas—all voiced their noisy histories, I could hear my part in it, could hear that the big hum is my collective stuff, both the noisy and the seemingly silent.

Thus a more thorough resolution to noise (and climate change) is a planetary stuff-diet: fewer tires rolling on pavement, fewer delivery trucks dropping off gear and gadgets, and fewer ships propelling through seawater bringing cheap clothes and china from China. And, quiet energy that is sun-powered and battery-stored (though those devices initially involve noise in their manufacture).

⌐ℓℓℓℓℓ

Urban planners say dense cities (where three-quarters of people in the developed world live) are needed for the planet's burgeoning population and unsustainable resource use. Some planners say it is possible to make cities more *biophilic*, in ways that highlight our connections to and love of nature. In his *Handbook of Biophilic City Planning and Design*, Timothy Beatley describes his ongoing work at the University of Virginia to conserve and restore the nature that already exists in cities, and to find or create new ways to insert new forms of nature. But the book's only nod to city "sound" ("noise" is not mentioned) is to "seek to integrate quiet areas" that restrict the mechanical and loud noises. He later notes a project at the Phipps Conservatory and Botanical Gardens in Pittsburgh that installed indoor "sound collages" of seasonal nature sounds because the triple-pane windows were a "blockade of nature" sounds from the outdoors.

While I applaud all efforts to incorporate authentic nature into cities, not acknowledging noise levels is a serious oversight. Our cities' collective racket is making all of us sick. Our hearts race and breathing speeds. We must talk louder, our voices hitting higher and higher pitches, our faces contorting, trying to communicate and claim our place in the landscape. Hearing loss is epidemic. Noise is a medical emergency, and dense cities require profound quieting. We need tranquility—not utter silence, but an acoustic ecotone that soothes our emotional brain and brings peacefulness and the chance to be physically and mentally healthy.

When BBC recordist Chris Watson asked people what sounds were tranquil to them, they said heartbeats, birdsong, footsteps, crickets, lapping waves, flowing streams. And sure enough, researchers tested such sounds and discovered they stimulated the limbic system in our brains that releases endorphins and feelings of serenity. Watson concluded that these natural sounds "are not soporifics but rather stimulating points of sonic light. They enable us to think clearly; there is a direct physical stimulus and a measurable clinical effect."

It is not enough to hear birdsong and flowing water piped through earbuds and sound systems; we need habitats in our everyday lives—wild and not-as-wild—that provide all creatures with enough acoustical space to be heard, to communicate without talking louder or singing at night. The complex symphony played in tranquil places is a key indicator not just of our physical and mental well-being, but of environmental quality generally, both for humans and the creatures with whom we share our spaces.

Another crucial step is ensuring that adults and children have quality experiences in quiet places. At camp when I was a teenager, two girls from a Chicago housing project who shared my tent complained that it was too quiet for them to sleep. After a couple of nights, sleep came easily (though they never stopped complaining about the bugs). Sometimes it takes familiarity with quiet to understand and truly hear the everyday noise.

Each Sunday in the city I seek tranquility in the nearby mountains, seek that base acoustical foundation upon which I can rest the week's burdens. One still, damp Halloween morning, my dog and I climbed the foothills at the city's edge for almost an hour. When we crested the high ridge and dropped down the other side, I could still hear it. Though it faded with each step down the leaf-littered trail, where stalks of thistles stood like tall gray ghosts, it was still there, humming. The loud sounds—train whistles, the beeps of backing trucks, motorcycles revving and shifting—had faded and blended into one big sound as we climbed. But I had not escaped the big hum of the big city I could no longer see.

The dog popped her head up, ears pricked. Suddenly—whop whop whop—a monster of a machine materialized over the ridge in front of us. A white helicopter rose and turned toward the university hospital at the edge of the foothills. The dog and I stared, silenced, watching the red cross on the machine's belly as it passed over us, cleared the ridge, and descended to the city, 110 decibels reverberating in our ears.

⊙ ⊙ ⊙

⊙ ⊙ ⊙

⊙ **6** ⊙

A Gardener Grows

"What a great backyard," he said, "but you've got a lot of clover in this corner."

"I know, isn't it great!" I said. "The grass didn't like the record hot summer last year, but the clover didn't mind one bit."

He looked puzzled but continued. "Well, we could take care of that for you. Besides lawn mowing, we also have a weed control program."

I laughed. That clover represented the beginning of my Freedom Lawn, a symbol of my transformed philosophy of what it means to be a gardener, a transformation that took me forty years. I have tended a yard (and often a garden) in ten cities across seven states, from the humid Midwest, to the wet Northwest, to hot and arid Salt Lake City. Gardening in those diverse habitats could have made me a master gardener, able to coax beauty and life from recalcitrant cultivars no matter the environmental conditions. Instead, over time, they challenged how I think about that great American duty of tending the land around one's house.

"No thanks, not interested," I said.

〰️

We all know the cultural icon of a good American yard: trimmed, deep green grass free of weeds or blemishes, smooth enough for croquet, admired during backyard barbecues. Dense riots of chromatic blooms from dozens of countries bursting in succession from spring to fall. Perhaps a "water feature" or a berm or a rock wall. Many people hire "lawn care companies" to create this seasonal performance by mowing, pruning, planting, fertilizing, and weeding; the rest of us spend weekends or summer evenings doing the same. A capital-G Gardener believes that with enough sweat, weekends, money, water, chemicals, tools, and perseverance, she can create this consummate "good yard." It's an American symbol of patriotism, packed with expectations for what it means to be a good neighbor. A good yard tells you responsible, hardworking people live there.

This exact same landscape is replicated from the Jersey Shore to Los Angeles and all points in between despite tremendously different conditions: rainfall, temperature, humidity, native plants, insects, and growing seasons. Americans have 63,000 square miles of lawns, which is about the size of Texas and several times larger than any irrigated crop. To achieve that perfect lawn, U.S. homeowners spend over a billion dollars *just* on outdoor chemicals, and almost a quarter of homeowners hire a company to spray chemicals for them. The chemicals applied to the land around our houses now far exceeds the amount applied on agricultural crops.

In a TV commercial for a lawn chemical, I watched a helicopter land on an expansive lawn and storm troopers bail out and race around the yard, each firing his spray tank on every living thing in sight. It fit the cultural stereotype that our yards and gardens are not places of cohabitation with other plants and animals but of all-out war against them. How sad. Here is a patch of earth you call your own, the place where you are most likely to encounter other life-forms. It's your touchstone of everyday nature and yours to tend and admire, and to learn so much from. It's your chance to be a lowercase-g gardener, learning how things grow and how animals live.

Some people—like my dad, like my friend Donna—find great solace and relaxation in tending this space; others find it drudgery. I have resided at every point in that spectrum. At this stage in my gardening journey, I am embracing my role as a lowercase-g gardener. I tend and care for my yard, but in large part, I am leaving the playing field and climbing into the stands to watch.

I can still picture my young self on my little knees in the garden, muttering about being sentenced to weed by Mom or Dad. The Iowa sun on my back and sweat running down my face, I plucked small shoots among the beets and beans. Mosquitoes buzzed about, finding bits of skin missed by the Off. Thinning carrots, especially when the soil was too moist or too dry, was sheer frustration, and I either pulled too many or broke tops from roots. Even more despised was mowing our one-third acre of grass. By the time the overnight dew had dried, the day was hot and humid. When the push mower ran out of gas, I'd gratefully head inside to stand in front of the window air conditioner, my clothes sweat-soaked down to my underwear.

Despite my loathing of garden chores, I remember equally the great joy of its bounty. Tomatoes still warm from the sun, corn picked just before it was eaten, zucchini, beet greens and tender beets. Pulling a carrot, wiping it on the grass to clear the dirt, and eating it on the spot. The rituals of preparation were sweet: sitting on the back step shucking corn with my brothers, and talking with Mom while we shelled peas, our conversation punctuated by the pop of pods and peas spilling into a bowl.

In childhood, leaf raking was a seasonal chore that was easy to turn into fun. My brothers and I jumped into and hid in towering piles of maple and hickory leaves, our sweatshirts and hair flecked with red, yellow, and brown leaf bits. In that era, we burned our leaves in the gravel road; we arrived at the dinner table smelling like we'd spent the day around a campfire.

Dad loved his yard and woods and relished changing into grubby gardening clothes to work outside. In early spring, he set up card tables in the dining room with trays of tomatoes and peppers he planted from seeds. He strode across the grass in a rain slicker and hat, spinning fertilizer from a hand spreader as the rain began, grinning. When spring frosts had passed, he planted impatiens and geraniums in the front planter, zinnia seeds over the septic tank, and outside the back door, begonias. Beyond the living room were big shrubs of pale hydrangeas and peonies and at the edge of the woods, a rose garden. Among the roses was a bronze plaque with a poem by Dorothy Frances Gurney: "The kiss of the sun for pardon, the song of the birds for mirth. One is nearer God's heart in a garden than anywhere else on earth."

My parents are gone now, and I can't ask them what the plaque meant to them. But given that I still remember its wording, it no doubt meant much to me. Since I am not religious in the God-sense, perhaps it reinforced that being outside was being close to the most important things—birds, sun, trees, and toads. I never felt alone outside, but surrounded and awed. Shafts of sun streaming through red maples. The damp smells rising after a midsummer thunderstorm. Owls hooting from the dark woods. The hush of fresh snow, newly tracked by rabbits.

What my Iowa childhood planted deep down were the tastes and smells of a yard's rewards and the seasons that ushered them along. It left indelible marks on me of how a loved and known place could provide great comfort and serenity. It also taught me the benefits not just of tending but of letting alone. My parents left mature, native trees standing when the foundation was poured. A stately shag-bark hickory tree smack dab in the middle of the gravel road was preserved, and cars went left or right around it. The lawn was watered only on the edge of death (in part because the water came from our well). Leaves and acorns were left where they fell in the woods that surrounded the yard. Our property was a landscape changed by humans to suit their desires, though by today's standards for lawn and garden care, it was hands-off light.

In my twenties and thirties I moved a lot, and the landlords of all those rentals were amazed at the effort I spent in the battered yards, planting gardens and bulbs, trimming, and tending. I am not entirely sure what spurred me since they were temporary residences. Perhaps I felt a need to belie the stereotype of renters having trashy yards, but it also just felt good to get on my knees, feel soil in my hands, and inhale all that sweet growth.

My most serious garden was in a tiny town in eastern Idaho that had a gigantic, flood-irrigated plot. The landlord next door would open the canal flood gate once a week, and we directed the water down deep furrows between the rows. The old root cellar between the house and garden was evidence that the patch had been tended for decades. I canned all summer: salsa, tomatoes, hot pepper jelly, pickles, dilly beans, pickled beets, corn relish.

Sometime in my thirties, Mom gave me a sweatshirt with the words "Plant Manager" emblazoned across a bouquet of bright, lush flowers. The moniker fit me then. I found great satisfaction transforming those neglected yards and abandoned gardens. I thoroughly enjoyed getting my body in contact with the stuff of the earth and getting out of my head. I felt part of something beyond my own species. And, the spruced-up yards provided pleasant places to read, eat meals, sit with friends, and celebrate the seasons.

As Plant Manager, I treated each rental yard as an interchangeable franchise in my gardening operation: Minnesota, Washington, Idaho, Utah, wherever. Even though I learned about the native vegetation that lay beyond the yard in each new region, I had been inculcated as to the exact flora that "belonged" around a house. And that's exactly what I recreated—especially the grass.

Though I didn't so label it, I was trying to create what writer Michael Pollan calls an Industrial Lawn. One, it's composed of grass species only—a monoculture. Two, it's free of weeds and other pests. Three, it's continuously green. And four, it's regularly mowed to a low, even height. To keep it thus requires considerable money, time, energy, and "inputs" (like fertilizer and chemicals). As Pollan noted, "It has the added virtue, at least in terms of the lawn care industry, of never being completely attainable."

Yet it is such an easy snare, thinking I control and manage that lawn and can achieve perfection. Even the act of mowing, with its intoxicating fresh-cut grass smell, brings a strange satisfaction that you have somehow restored order with those trimmed blades of grass, that you rendered wild nature once again fit for human habitation.

⁓ᵒᵍᵍᵉᵇᵇ

I bought my first house when I was thirty-nine, a year after graduate school in Minnesota and with a new job at the university in Salt Lake City. The tiny one-bedroom 1917 bungalow was surrounded by green grass, and in the backyard were a couple of lilac bushes, some rose bushes, and an apple tree. I was thrilled with the possibilities.

I trotted home from the garden center with my Minnesota favorites: forget-me-nots, hosta, and impatiens. My friend Camille gave me as a housewarming present one of her favorite mountain flowers, a lupine. Snails devoured most of the hostas the first night, the culprit revealed by slime trails the next morning. The lupine was entirely gone; only the plastic stake marked the spot. Despite watering, the forget-me-nots and impatiens succumbed to the intense dry heat. For several years, I was a slave to my hoses, watering something—grass, plants, flowers—every damn day. St. Paul and Salt Lake couldn't be more different in terms of humidity, precipitation, and temperature, yet I fell into the trap of thinking I could make these habitats interchangeable with enough Gardening.

That experience began a long, slow process of altering how I felt about my Plant Manager role and breaking down the belief that yards should consist of the same general flora (geraniums to marigolds) regardless of highly specific habitats and climes.

Over several summers, I dug up all the Kentucky bluegrass on my tenth-acre and replaced it with plugs of buffalo grass, a drought-tolerant native that eventually grew into a lush mat about four to six inches tall with tops bent over slightly; it required no mowing! Gradually, I replaced the thirsty shrubs and flowers for natives that only sipped water.

Nevertheless at times, I felt engaged in the battle of Sisyphus, a king of Corinth in Greek mythology whose punishment in Hades was to push a large rock uphill, only to have it roll back down and be forced to begin again. Trying to make the yard "natural" when it was surrounded by exotics and invasives seemed hopeless, as if I needed to take part in this battle every dang year and would so every year hence.

It did not help that my two adjoining neighbors had gardening styles at opposite ends of the spectrum. To the west, Bob spent countless hours on his yard, digging dandelions as soon as they bloomed, constantly adding new features and plants, and watering, watering, watering, even during a stiff breeze in the hottest part of the afternoon. To the east, people walking past Peter's yard asked me if the house was vacant. He did cut the grass with a push mower and watered occasionally, but otherwise he never raked a leaf, pulled a weed, or trimmed a bush—which meant that the leaves, weeds, and bushes liked to travel to my place. I admired parts of Peter's au naturel approach, but in such a patchwork of small lots, his approach became my problem. Soon, I battled bindweed and vinca arriving from Peter's, and Kentucky bluegrass seeds arriving on the wind and rooting in the buffalo grass.

The invaders should not surprise me: polyculture is inevitable! My buffalo grass may have been a native grass, but it was still a monoculture, just like the bluegrass lawn it replaced. In a monoculture there will always be opportunistic plants—which are taller, shorter, less thirsty, adapted to disturbance, or more tolerant—seeking to join in. Biodiversity is what provides the best resistance to noxious species invasion.

To a casual observer, yards belie the effort it takes to produce them. In reality, a city is one gigantic disturbance zone that would revert to thistles, dandelions, and bindweed if residents didn't engage in a ceaseless, continuous battle (typically with chemicals). As a landscape, a city has been too long altered and disturbed, and it's a crazy hodgepodge of flora yard to yard, making it difficult to be a less-vigilant gardener. The heavily watered yards have even altered Salt Lake City's microclimate, increasing humidity and attracting insects and plants suitable to the altered conditions.

What keeps a block of lawns in check (in addition to city ordinances) is palpable social pressure. A survey of Virginia lawn owners found that 80 percent thought their lawn was average or below-average, and they weren't satisfied with its present condition. As Paul Robbins concludes in *Lawn People*, our lawns make us anxious. He found that lawn anxiety—not wanting to feel out of place or insult the neighbors, worrying about social status and property values—results in the massive use of lawn chemicals. In many ways, those "just treated" flags stuck in the grass signal not environmental danger but social reward and pride: see, I've done my part for the neighborhood. The social pressure is so intense, Robbins found, that lawn chemical users (who tend to be wealthier and educated, and surrounded by neighbors who use chemicals) tend to be *more* worried about chemical usage than those who do not use them. Yet they use them anyway; that's serious peer pressure. All to achieve the never completely achievable Industrial Lawn.

The "stay off until dry" lawn flags might make us feel better about their use, but in fact lawn chemicals drift in the wind, volatilize, leach downward, and more important, catch a ride on human hosts and other animals. Studies of household dust in areas where people use lawn chemicals are disturbing; the chemicals are released from clothing, shoes, and pets onto carpets and countertops—even if you do not use them on your own yard. Family pesticide use (including lawn herbicide application) has been shown to be related to childhood brain cancer. Other case-control studies of malignancies in children that are linked to pesticides include leukemia, neuroblastoma, Ewing's sarcoma, non-Hodgkin's lymphoma, and cancers of colorectum and testes. Robbins says, "In sum, these chemicals by no means simply stay where they are put and vanish when they are done. Nor should we be surprised. This is the first rule of ecology: nothing goes away." I used to move with my dog from the sidewalk to the street when I saw a just-treated sign, but I now realize that this location is not much safer.

Despite these documented dangers, we view the management of our grassy piece not as a choice but a moral and cultural obligation, and we feel the heavy burden of the normative power of the community. Marketing from lawn care companies likewise reinforces your duty to your turfgrass. This multipronged manipulation creates anxious "turfgrass subjects." But, as Robbins concludes, "Anxious subjects need not be docile ones."

~~~~~

I got my chance to step down as Plant Manager when I bought ten acres of land in Wyoming with plans to eventually build a cabin. With a nine-month academic contract, I could easily transplant myself to Wyoming for the entire summer. Thus, I had to make my Salt Lake yard as self-sustaining as possible; all the vegetation was now drought-tolerant but still required some watering. I dug the trenches for a sprinkler system and got it running before I left town.

In the summers I lived at the cabin, I increasingly acquiesced maintenance of the Salt Lake yard to the plants themselves. Through the plants' survival (or not) and their movement, they educated me on what they needed. The hummingbird vine was so happy it attempted a takeover of the utility pole (which the utility people did not appreciate). The pink penstemon didn't really want to "bloom where it was planted" and moved itself to other locations. It seemed to like companions and took root under the rabbitbrush, next to the Indian ricegrass, and tucked under the rim of the bench. All these places were near the sand path, so perhaps it was a drainage or soil issue. The Utah chokecherry tree wanted to start an entire colony and sent shoots for new trees up to ten feet away (much to Bob's ire), and its growth soon overshaded the once-sizable purple coneflower beneath it. Another coneflower in the front yard was both too shaded and too wet, thanks to Bob's ultramoist yard.

Once I became less attached and more absent from the city yard, it was a relief to become more of an observer and a less active participant in their little plant lives. Thinking I could (or should even try to) manage their lives was like thinking I could manage the rain. It was liberating for us all.

Much of my early spring yard work in Salt Lake was deadheading last year's growth, which filled an entire yard-waste bin. As I wheeled the bin to the curb one spring, I realized that it would never occur to me to deadhead in Wyoming! When I arrived there in late May, the new fuzzy shoots of balsamroot poked up through last year's leaves, which splayed in a pressed and dried ring around the new shoots. Grasses flattened and matted down by snows had new growth poking up just fine from underneath. Such old growth is nature's mulch and fertilizer. So why didn't I leave it alone in the city? Why was I attached to the neat and manicured look in my city yard when it was unthinkable and out of place in Wyoming? Yet year after year, I could not leave it alone. It would look too messy, too unkempt; no good gardener could leave this be.

Some cabin neighbors had seeded lawns, which they mowed and even watered, but I did not want a yard or flower boxes or baskets of annuals to tend. I wanted a seamless line from cabin to meadow and woods, a place that "fit in" to the greatest extent possible with its surroundings. I wanted to step to the background and pay attention to what bloomed where and then mimic that.

For three autumns I seeded the construction zone, beginning with the steepest pitches near the cabin. I identified the grasses and flowers growing nearby and ordered these seeds from a business that specialized in Rocky Mountain flora. After raking, sowing, raking, and compressing the soil with hundreds of my baby steps, I covered it with rolls of wood shavings pressed between photodegradable green plastic mesh. It worked for the most part. Occasionally, a dog paw or deer hoof would pull up a corner, despite my anchors of rocks. But the cover held the seed in place and preserved moisture when summer came.

A primary reason to reseed these areas was to forestall weeds that jump at the opportunity of bare soil—just as they do in disturbed areas such as roadcuts. Nevertheless, what grew best (and first) in my reseeded areas were these uninvited guests, the opportunists who were phenomenal at taking advantage of such a situation. Thistles, roadside clover, dandelions, and cheatgrass. I pulled the worst offenders until the native grasses could better compete.

I truly enjoyed this project, but I didn't feel like a gardener— more like a restoration ecologist. I tried to place my desires aside and to restore in some sense what I had put asunder. It was humbling and a constant challenge discerning where or how the natives liked to bloom. I had no idea why the yarrow was so thick in one area and why the flax took hold in another. Or why the ground remained bare in several spots after several years, while the bunch-grasses went wild in what I considered unlikely places. The paintbrush seed I planted never germinated, nor fireweed nor penstemon. They all grew elsewhere on my land but wouldn't deign to sprout where I planted them.

Thank goodness that regardless of all my efforts, all the naturally located native grass, flowers, and trees still sprouted, grew, and died. After all, these plants took eons to get it right, to find just the right magical combination of conditions—soil, sun, slope, moisture, temperature, wind, and neighbors—to root and bloom. It made me wonder what Salt Lake yards might look like today if pioneers had arrived not with visions of European gardens but with a desire to fit in with the natives.

On a walk from the cabin onto Forest Service land along a little creek, I spied some potentilla with their deep yellow flowers and frilly green foliage (also called shrubby cinquefoil). Perhaps I could plant a couple of these shrubs around the cabin; the species is popular in nurseries and I see it in Salt Lake yards. When walking home, I realized why I had never seen any growing near the cabin. Duh: they liked to keep their feet wet! They, like every other shrub, tree, grass, and flower, knew best where to plant themselves. Of course, I am free to try and rearrange that to suit my desires, and I may get lucky. But in all likelihood, it will take more effort and tending by the Gardener.

Capital-G Gardeners are able to make a yard look the same each and every summer, regardless of what the weather and season deliver. I did not realize this until summers five and six at the cabin, when I watched but didn't intervene in two very different seasons.

By late June in the fifth summer, forest fires burned and meadow grasses had gone to seed (and I began to mow a large, broad circle around the cabin for "defensible space"). Some early bloomers made an appearance; many others parched on the vine. I mourned the desiccated meadow as flower tops faded and stems turned crisp. I missed the blooms, and it seemed to signal a quicker passing of summer. Its length was unchanged of course, but I felt pangs of sadness, remnants of city life that say all growing things should be green, and if they're not, you should water them.

But ah, in summer number six an excellent snowpack and a wet June produced a riotous progression of blooms. It looked like bloody Switzerland, with mounds of early mountain bluebells, then carpets of yellow—first balsamroot, then arnica—followed by lupine, buckwheat, and penstemon. The dry slope that held at best a dozen sego lilies erupted with them. Lavender harebells propagated happily. For the first time, red was more than a wee accent when scarlet gilia produced an amazing display. By August, despite six rainless weeks, the flowers bloomed.

In retrospect, my distress with that dry summer was very curious (and highly cultural). Underneath the parched surface, down with the moles and salamanders, mixed with the scat of hoofed visitors, lay the dormant seeds, the roots, the stems, all waiting. For them, it wasn't a matter of patience or relinquishing control— just a simple accommodation of cycles, of taking the soft path of seasons, of living with what is.

For the most part, the therapy I once got from gardening morphed at the cabin into watching a wild garden do what it knows best. Nearly every day, I strolled through it to see who was blooming (ah, the first leopard lilies), who was seeding (look, the purple clematis have turned to white wisps), and who was being invaded by pushy neighbors (collomia annuals have taken over the septic field). I wandered and watched. No deadheading blooms, no lawn, no watering. I was marvelously unneeded.

~oggopp~

Then, I bought a new house in Salt Lake. I had looked for a year, wanting something closer to campus with room for an office and guests. It was a charming 1939 stucco cottage with a huge yard. I bought the house in February, knowing it had housed renters and clueless about what lay under the snow.

About seventy-five years ago, the entire neighborhood was orchards and farmland with good soil on a bench above the deep gully of Parley's Creek. A 1939 aerial photo of what is now my neighborhood showed only clusters of treetops, rows of crops, and a barn. I tried to keep this picture in mind when the scope of my outdoor task loomed large. And I remembered that a very lumpy yard is a good thing; my friend Flo said it means there are lots of worms.

The entire quarter-acre yard is grass, though in the backyard, the back third has only sparse crabgrass and a half dozen shrubs. It's shrubby and messy, and it has potential. Rimming the driveway are old-fashioned lilacs and honeysuckle.

The whole bloody yard is full of bindweed (a perennial

morning glory), spiraling up a shabby rose bush, ensnaring shrubs, and permeating the lawn. From experience at my first Salt Lake house, I know that bindweed would likely survive nuclear war; it is that resilient. Its thick white knobby roots hang out deep below the sod, not minding if the aboveground growth is pulled or sprayed. Also growing in the grass are clover, creeping jenny, dandelions, and a half dozen other "weeds" I recognize but can't name.

But this gardener has grown; I now know what is not worth the fight. I would rather watch and enjoy my yard than battle. I'm calling it my Freedom Lawn (Pollan's alternative to an Industrial Lawn), so named because it encourages a hardy polyculture where all kinds of plants grow, both those that planted themselves (clover, violets, and dandelions) and ones I might plant (such as water-sipping perennial grasses). A Freedom Lawn is more adaptable and tolerant of various kinds of stress (thus less disease or insects) and can require less water. The clover that the lawn care guy wanted to poison was evidence that my Freedom Lawn had begun, all by itself.

My first lawn-reduction project was two concrete patios in the backyard: one to connect the house with the garage, the other in a parched sunny spot choked with weeds. Then I cut the sod from a large swath behind the garage. There and in the abandoned garden space in the back-third, I envisioned a wild-ish woods. I planted a dozen Gambel oak, the scrubby oak in the foothills surrounding the city that can grow taller with some watering and lacks lower branches, producing an open wooded feel.

My first attempt to kill the crabgrass in the abandoned back-third was to smother it with leaves. I piled all my leaves (and those of several neighbors) onto the area, stomping and crushing the leaves into a thick matt. But in spring, the crabgrass poked through, ecstatic for the new moisture-preserving mulch. Hah, what do I know! My second attempt was a strip of heavy-duty landscape fabric with holes cut for the trees. A friend and I made several trips to the dump with his pickup for wood chips, which we put on the fabric. A sprinkler guy plumbed a new drip system, and I attached feeder lines to each little tree. Beyond the weed fabric, the crabgrass and weeds still thrived, until the shrubs and ground-cover I planted there are able to better compete them, I weed-trim.

Most recently, I hired a crew to remove the grass from the parking strip (between sidewalk and street) and replace it with light-colored large gravel over weed fabric. Around each tree, I planted a ring of perennial bunchgrass. I know this violated the social norms of the block (where grass grows in every parking strip) because I similarly violated them at my first house. There, while I was removing grass in the parking strip and replacing it with creeping thyme and bunchgrasses, people stopped and looked, and some asked questions. Now, over half the parking strips on that block are grass free.

People ask me what I do with my yard when I'm in Wyoming all summer. I tell them I have a sprinkler system, and I hire students to mow every two weeks. When I visit once mid-summer, I weed-trim along the back fence and otherwise do some cleanup. That's all I do.

I do feel social pressure from some neighbors and friends for the state of my Freedom Lawn. Most of the yards on my block are deep green, weed free, and sprout "just treated" signs each month. The friend who waters my houseplants in summer sent me a photo to point out the weeds in my yard. One neighbor texted me that my grass was looking pretty brown. I thought (briefly) about planting my own lawn sign in response to the "just treated" signs: "Proud Owner of a Freedom Lawn." Or, I could proselytize my small-g gardener philosophy to anyone who would listen. But in the end, I just trust that the process (with my lack of anxiety and abundance of freedom) speaks for itself, even if it bucks a pro-lawn, pro-chemical culture.

As I have grown as a gardener, so has my city. Many of the native plants I mail-ordered twenty years ago can be bought locally now. The modest weekly farmers' market has exploded, as have cooperative and community gardening and backyard chicken coops. Though my block has no xeriscaped yards (from the Greek *xeric*, or scant moisture), they are sprouting throughout the city. Local, native, and drought tolerant are now part of urban parlance. However, many gardeners are learning through experience that "zeroscaping" (a common mispronunciation and misunderstanding) does not mean planting natives and walking away; it takes a lot of tending to establish even hardy plants, and it will forever (yes, forever) need weeding and some water.

My golden retriever loves our backyard, which I take as a good sign that I have grown as a gardener. She loves rolling on the big patch of (freedom) grass that remains, where a huge silver maple towers in the center. She sleeps on the grass in the sun until she gets too hot and retreats to the new concrete in the deep shade. At the shrubby, messy back edge of the lot she sits and stalks quail, and watches the family of rodents scurry from their burrow in the hollow stump just on the other side of the fence. She likes to chew on the crabgrass that pokes out at the edge of the weed-block fabric (a good reminder of why I don't spray it). The branches that fall from the back maple trees and quince are great chew toys. She enjoys watching jays, finches, and juncos at the feeder and helps cleans up the seeds they drop.

⁓ღფიდ

Thus far in my quest to understand what it means to be a gardener, here are some things I have learned.

There is much less stress and pressure in my life if my personal identity and patriotism are not tied to how my yard looks.

Love and admiration for all things that grow and breathe on the patch of ground around my house are a crucial connection to the nature in which I reside, which is connected to (and has consequences for) all nature.

Accept clover, dandelions, insects, and animals who are likewise looking to make a living and sustain themselves; it's much less work and stress than engaging in constant battle.

Chemicals do not stay where you put them. What goes on, comes around.

Gardening is more about observing and learning from plants, soil, and creatures of all kinds than about attempting to control and manage them. And it's a relief not to be in charge.

○ ○ ○

○ ○ ○

○ 7 ○

Speaking a Shared Language

"When I use a word, it means just what I choose
it to mean, neither more nor less."

—HUMPTY DUMPTY
(*Through the Looking-Glass*, Lewis Carroll, 1872)

Humpty Dumpty was way wrong; words are not so uncomplicated. As David Orr, a well-known expert in environmental education, said: "Words have power. They can enliven or deaden, elevate or degrade, but they are never neutral because they affect our perception and ultimately our behavior." That power includes how language presents everyday nature to us: "weed," "eco-friendly," "park."

When I open my mouth or type something on the keyboard, I'm premeditating only superficially about the words. The language that comes out is more than a simple vehicle for conveying information; it is a conceptual framework for meaning making that reveals unconscious (and conscious) cultural assumptions and worldviews. It directs attention and limits a field of view.

Language choices shape us, the research says, beginning in childhood. What interests me is how the language of *anthropocentrism*, a strongly human-centered view of nature, shapes and disconnects us and reinforces a hierarchy of humans over nature. Of course, we have no choice but to use a human lens to view what's around us. But there's a substantial difference between seeing through human lenses and believing that humans are the only species among millions that really matters. Anthropocentrism sees humans as somehow outside of ecology and independent from all the linked systems of the planet and magically beyond their influence. Anthropocentrism sees the "beyond-human"—other animals, plants, air, water, minerals, landforms, and myriad ecosystems—as mere instruments to satisfy immediate human desires.

A man who made a delivery to my cabin in Wyoming scanned the forest floor in several directions, then scanned the treetops.

"You got a lot of dead timber here going to waste," he said. "Want me to log 'em"?"

I explained that dead trees were valuable to many species like woodpeckers and provided organic material to the soil. Mid-sentence, he waved his hand dismissively and left.

How human-centered language disconnects us from nature is illustrated in four vignettes that follow. For each, I rewrite the language in a way that envisions what more connected language might look and sound like. But first, a linguistic look at anthropocentrism and English.

~~~

When you ask people about the roots of our environmental crises, many point to anthropocentric beliefs. Yet some claim there are linguistic roots to the crises as well. After all, the language we use and the stories we tell about ourselves and the world around us reflect cultural values, priorities, and relationships.

A graduate student asked me for advice on the best language to refer to all that is not human. "If I call it 'nonhuman nature,'" he said, "that implies we aren't part of nature, which of course we are, so is 'beyond human' better? Or, should I use 'environment' instead of 'nature' because the concept of 'nature' is such a social construction? But 'environment' seems so sterile and political. And if I say 'animals,' how do I make it clear that I'm including humans in there?"

I too stumble over the "correct" language, which academics and others seem to make worse even in our attempts to make it better. Farmer and author Wendell Berry goes a step further: "We are using the wrong language . . . We have a lot of genuinely concerned people calling upon us to 'save' a world which their language simultaneously reduces to an assemblage of perfectly featureless and dispirited 'ecosystems,' 'organisms,' 'environments,' 'mechanisms,' and the like. It is impossible to prefigure the salvation of the world in the same language by which the world has been dismembered and defaced."

The English language separates and "dismembers" humans from nature in ways that other languages do not. According to biologist–turned–social scientist Brendon Larson, English in particular amplifies the distinction between nature and culture. In Western societies, we inhabit a subject-object frame, where humans are the subject and nature is the passive object.

The languages of many indigenous peoples instead relate to nature within a subject-subject frame, which emphasizes being in relation. When Quebec researchers wanted to understand people's perceptions of the concept "wilderness," they interviewed Cree Indians, Vermonters, and others. The Cree had no word or concept comparable to "wilderness"; they said it made little sense to separate themselves from it. Instead, they associated this "wilderness" with a state of mind and described it in much more personal ways. In contrast, Vermonters in the study defined wilderness as the absence of humans and human disturbance, and they saw no contradiction between the "unpeopled" part of their definition and their desire to visit it. Others viewed wilderness as places that were valuable for potential resources or as escapes.

Since languages evolve over time, has English become more environmentally friendly and connected to nature? After all, "sustainability" and "renewables" are now everyday vocabulary. It depends on the words lost, words gained, and the worldview behind the words.

In a new cafe on campus, I stood at the counter for several minutes until a young female employee pointed to a screen I had passed.

"You order there," she said.

After I made my selection, the screen asked me to choose "To Go" or "Stay."

I asked her, "If I choose 'Stay,' does that mean I'll get my food on dishes?"

She stared at me.

"What do you mean 'dishes'?" she finally asked.

"You know, real plates." Still a blank stare.

"Ceramic plates, you know, that you wash and reuse. Dishes."

"Oh, no, I don't think so."

What was disturbing about the exchange was that *dishes*, these environmentally friendly things, were missing from her lexicon; disposables were her dishware frame of reference. For me, the hole in her vocabulary was just as serious as the environmental practice. I recognize that cultural usage makes language fluid; we add words like *drone* and *texting*, but what words disappear? Since pioneer times, U.S. English has "lost" about a thousand words.

The *Oxford Junior Dictionary* added the words *analogue, bullet-point, celebrity, voicemail,* and *bandwidth*, and removed the words *acorn, willow, lark, minnow, clover, buttercup,* and *magpie*. During the vociferous public response to the change, author Tom Shippey said, "If you have no vocabulary for things, you notice them less. So you start thinking they're not important, and we lose touch." Removing words that connect us to nature and demonstrate our understanding of it devalues our language and our lives.

Now, four vignettes of disconnected language.

Bugzillas!

At the end of my block, a long banner hanging from the street lamp exclaims *Bugzillas!* Above the blood-red letters, a goliath black-horned beast of a beetle, rimmed in dark red with red bulging eyes, towers above a crowd of tiny people. Some people cower with their arms in front of their faces, and others are running away. Behind the beetle is the tower of the local Union Pacific train depot with an American flag. Above the building, two fighter jets streak toward the beetle.

Two street lamps down is another *Bugzillas!* banner with a building-sized green praying mantis (also with red bulging eyes) against a cityscape. Its front spiked appendages hover over crowds of people fleeing in terror. On the next banner against a red background, a behemoth-sized black, hairy spider waves its legs above the terrified masses, with two sharp "daggers" just above their heads.

This was not a Hollywood remake of Godzilla with invertebrates, but advertising for the newest exhibit at Utah's Hogle Zoo.

When I returned home from my walk, I visited the zoo website: "Have you heard the buzz? Giant bugs have invaded Hogle
Zoo this summer! It's an infestation celebration and it's happening this summer only at Utah's Hogle Zoo! BUGZILLAS! presented by the Les Schwab Tire Centers, features 14 incredible,
larger-than-life animatronic bug species all located throughout
Zoo grounds."

Yet, the rest of the text made the ferocious and fearful giants
seem, well, likeable. It called the "bugs" sorely misunderstood
and taken for granted, bugs that were helpful pollinators, garbage collectors, soil conditioners, and natural fertilizer producers. It said, "Insects are the foundation of all life on earth. Without them, mankind would simply cease to exist." These are the
same Bugzillas?

The zoo's community relations and media person told me that
the banners and TV commercials were designed to get people in
the door, and once they were "on ground" they would receive the
conservation message. She said the zoo's ad agency designed the
banners so people would see the bugs as fake and really big, and
that parents would see the tie-in to Godzilla. She said the accompanying TV commercials were filmed in shaky black-and-white
and showed big bugs crawling over Salt Lake City.

Hogle Zoo (like other zoos) likes to tout their education and
conservation mission, yet the language of Bugzillas! shouts entertainment. The language distances us from the most predominant
of the world's living creatures, and amps up (sadly familiar) fear
about "bugs." The Bugzillas! story is one of enemy invertebrates
that humans must "shoot down," a message reinforced by the cultural icon Godzilla. Website words that embrace bugs as friends
and allies in our everyday lives are lost in the din.

How would I rewrite *Bugzillas? Super-heroes who run the world!* In
the banners and TV ads I would show adults and children holding
or gazing lovingly at invertebrates. And I would fill the zoo with
actual-sized living insects, spiders, worms, and slugs—their sheer
numbers make them huge, no giant robots required (well okay, I'd

supply some magnifiers). Invertebrates are not objects and ene-mies, but community and subjects in their own right.

Cuties

The blue mesh bag labeled them "Cuties." In a drawing on the bag, a citrus sphere sat atop several green leaves; a half-open zipper on the orange peel revealed inside a face with whimsical human eyes, eyebrows, and a cheeky smile. Advertising chose to anthro-pomorphize the fruit and give it human qualities to show kids (and moms) that this mandarin orange is human friendly because it's easy to peel. The image got the shape wrong: mandarins (and tangerines and satsumas) are oblate, not spherical.

I browsed the official Cuties website, where every page fea-tured Claymation-type farmers, children, and families picnick-ing beneath the orange groves in California. "Welcome to Cuties Country" with expansive blue skies, puffy clouds, and butterflies. Male Claymation farmers in wide-brimmed straw hats climbed old-fashioned ladders to pick the oranges, with nary a machine in sight. Cuties were family owned and family grown: "Grown with love for those you love."

Anyone familiar with modern industrial agriculture knows this is a storybook version. Just like a grocery ad for beef with a cowboy riding the range at sunset amidst his herd, the language of food production is highly sterilized and idealized. Gone are the machines, the crowded (and often inhumane) conditions, the chemicals, the migrant farm workers, the prodigious water and land use.

I bought this bag of Cuties for a class experiment: to see if stu-dents might be able to shift their perceptions away from human centered and object oriented through mindful focusing. First, I asked them to close their eyes and take twenty slow, deep breaths while I placed a mandarin on each desk. Then, I asked them to put all of their focus just on that orange and to use every one of their senses while they slowly peeled and ate the fruit.

When I asked them to share their experiences, the first volun-teer said, "That was the most awkward thing I've ever done," and we all laughed. Then the comments flowed.

"I kept thinking about why the texture on its skin was wrinkly like that. I knew there had to be a good reason but I didn't know what it was."

"I was just amazed at all the juice, and even the sound of it in my mouth. I don't know, I just never really noticed all the juice before."

"I kept thinking about all the interdependencies. Like after I eat it, what will happen to the peel? Kinda like when I go hunting, I leave the innards and I know that other animals will use it and benefit from it. So I wondered, if this fruit were growing in the wild, what animals would eat the ones that fell?"

"I kept looking at all the little membranes that I never really looked at before and wondering why they were there, like what they did."

"I don't think I've ever used all five senses when eating something. I probably won't ever look at an orange in the same way."

"Okay, so this is gonna sound strange, but it was like I was eating a living being."

Many students were successful—not only in relaxing and mindfully eating, but in shifting their view of the mandarin from Object to Subject. They thought about the orange's structure and life, the interdependencies with its environ, and the interaction between person and fruit.

It is more difficult in English to see an orange or an insect as a subject, because in English every other living being (animate or not) is labeled as an "it." Only humans can be a *she*, a *he*, or a *who*. When the students considered the fruit thoroughly, language considered it as more of a being in its own right; the mandarin had *animacy*.

If you have studied French, German, or Spanish, you are familiar with the gendered articles like German's *der* (masculine), *die* (feminine) and *das* (neuter). Do gendered articles make a difference in how you perceive something? According to one study, yes. When Germans were asked to describe a "key" (which has a masculine article in German), they described a key as hard, heavy, jagged, and serrated; speakers of Spanish (where "key" is feminine) described "her" as golden, intricate, lovely, shiny, and tiny. A "bridge" to a German (which has a feminine article) is beautiful, elegant, fragile, peaceful, and slender. To Spanish speakers, bridges (with a masculine article) were big, dangerous, strong, sturdy, and towering. By extension, all the nonhuman *its* in English have neutral or just plain impersonal characteristics without animacy.

Puhpowee. In the Potawatomi language, this word translates to "the force which causes mushrooms to push up from the earth overnight." Essentially, the life force of mushrooms. The verb *puhpowee* became a signpost for Robin Wall Kimmerer, a botany professor, member of the Citizen Potawatomi Nation, and author of *Braiding Sweetgrass:* "The makers of this word understood a world of being, full of unseen energies that animate everything." The Potawatomi language embodies a *grammar of animacy* that maps relationships among all beings on the planet.

And it's not just mushrooms with a life force. Seventy percent of Potawatomi words are *verbs*, compared to just 30 percent in English. (The Blackfeet Indian language Niitsi'powahsin has a similar preponderance of verbs.) As Kimmerer writes, "The language reminds us, in every sentence, of our kinship with all of the animate world." Potawatomi does not divide the world into masculine and feminine, nor hold *it* separate as in English. She says, "The arrogance of English is that the only way to be animate, to be worthy of respect and moral concern, is to be a human."

For days after reading this, I walked around, imagining the objects around me as animate subjects *who* interacted with me in constant fluxing relations: the flower who smelled so sweet, the wind who was so alive, the trees who gave me this book. It shook me, this communication professor who just learned that the language of her culture (and much of the world) was its own iron cage for how we see and value the world. My noun-based English codifies and reproduces a human-centered culture obsessed with things and not being.

Kimmerer struggles to learn her native, fast-disappearing language. She reimagines English's static nouns as beings and conjugated as verbs in Potawatomi:

A bay is a noun only if water is *dead*. When bay is a noun, it is de-fined by humans, trapped between its shores and contained by the word. But the verb wiikwegamaa—to *be* a bay—releases the water from its bondage and lets it live. . . . To be a hill, to be a sandy beach, to be a Saturday, all are possible verbs in a world where everything is alive. Water, land, and even a day, the language a mirror for seeing the animacy of the world, the life that pulses through all things, through pines and nuthatches and mushrooms.

If the maple tree in my backyard is a *who* and not an *it*, "know-ing" it is an entirely different enterprise. Knowing is more than just learning the facts of its natural history—the texture of its bark or the feathery blossoms springing from grainy red buds. Knowing the tree as a somebody, a member of my immediate community, a fellow being on the living land feels good—and right. Kimmerer hopes "a grammar of animacy could lead us to whole new ways of living in the world, other species a sovereign people, a world with a democracy of species, not a tyranny of one."

In an essay "Speaking of Nature," Kimmerer explained how "linguistic imperialism" has always been a tool of colonization; native languages were the enemy and native children forced into boarding schools were forbidden to speak them. Their language was an affront to colonists, she said, because it is "a language that challenges the fundamental tenets of Western thinking— that humans alone are possessed of rights and all the rest of the living world exists for human use." English became the dominant language of commerce, and the forest and copper ore became equivalents *its*.

It's not surprising that my Cuties rewrite sounds utterly for-eign to modern advertising:

To be a Mandarin Orange is a great journey. Our home was Southeast Asia. Humans colonized us in the United States by 1900, although humans mainly shipped us here from Japan as a holiday treat. On the land, we live in a community that nur-tures us—the bees who pollinate our blossoms, the worms and grubs who enrich soil for our roots, the waters who feed us, and the energy of the sun who makes our lives possible. Be grateful for the gift of each orange and eat it mindfully. If you do not eat the peel or use it in cooking, give it back to the Earth so she can return our nutrients.

Wolves and Resources

Wolves: Government-sponsored Terrorists. I saw this bumper sticker on a Ford pickup truck in a Dollar Store parking lot in Wyoming. The chosen words speak to larger patterns of thought, such as the metaphors we use to describe (and justify) *resource management.*

A metaphor is an implicit (and often unquestioned) comparison: "time is money," and we "waste," "invest," and "spend" both. Social scientist Larson explains that metaphors enable us to understand one thing in terms of another, and to think of something abstract or novel in more concrete terms. If you *harvest* trees (or deer), that implies they are a crop that humans grow, and this comparison catalyzes actions to produce the biggest crop.

Managing evokes a culture of control by experts (who are of course humans). The dominant managerial approach is mechanistic and reductionist: a lawn management company reduces your problems to "weeds" and "pests" (which are mechanically and chemically removed) and to "nutrients" your lawn may be lacking. Your yard is an expansive passive object, not a community of beings. Subordinating the nonhuman makes it easier to justify their exploitation.

Anthropocentrism holds that humans can and do *control* nature. Nature is the machine and humans the engineers. Over the last century, humans "controlled" wolves out of existence, for they competed with humans' livestock crops. Wolf reintroduction represented relinquishing any sense of control over these predators, and what's more out of control and dangerous than terrorism. (The word *terrorist* is tossed out frequently these days. A "terrorist" engages in "unlawful violence and intimidation," which applies aptly only to humans.) A cultural assumption in the bumper sticker language is that the public lands where cattle and sheep graze are best used for these commodities, not for the creatures who consider this place home. Another anthropocentric assumption is that the government can and should attempt control of a single element of an ecosystem to favor human commerce.

Natural resource management. When I worked for the Bureau of Land Management, those words tumbled easily off my tongue and onto the typed page, as though their meaning was (as Humpty Dumpty said) exactly what I chose it to mean. I also wrote of *range improvements*, which were improvements for cows and sheep, not ecosystems. *Range* is its own metaphor because it labels land not by ecological land type but according to its human use (livestock grazing). The metaphors I used disconnected rather than integrated humans with nature and made humans the active subject over a passive object nature. I promoted in news releases and planning documents that human managers were the official knowers and doers who operated with scientific facts that were cleanly divorced from the values of human culture. I now see the fantasy that humans were ever in charge of these earth elements.

A *resource* is defined as "something that lies ready for use" and *resources* as "wealth and assets." Larson explains how English words promote environmental exploitation in ways that appear neutral (*development* and *resource*), or positive for unpleasant things (*culling* and *improvements*), or pejorative (*overmature* trees). If we speak of nature as a *resource,* that is the way we are likely to act toward it and value it. This did not occur to me, really, when I was enmeshed in a "natural resource management" job.

Etched in limestone on the engineering building at the University of Wyoming are the words, "Strive on—the control of Nature is won, not given." The inscription prompted John McPhee to write *The Control of Nature,* a trilogy that explores the all-out battles and extraordinary lengths to which people go to attempt to control nature, a "struggle against natural forces—heroic or venal, rash or well advised—when human beings conscript themselves to fight against the earth, to take what is not given, to rout the destroying enemy, to surround the base of Mt. Olympus demanding and expecting the surrender of the gods."

In Iceland, McPhee observed a physicist who attempted to direct molten lava away from town. In the Mississippi delta, the Army Corp of Engineers built a great fortress to restrain the flow of the Atchafalaya River and compel the Mississippi not to change direction. And in Los Angeles, McPhee chronicled how the city built (at extraordinary expense) 150 stadium-like basins in an attempt to catch material from debris flows near LA's famous canyons—flows that pluck up trees and cars, burst through doors and windows, and fill houses to their eaves.

If you believe in your mastery of nature, it follows that you are likely ignorant of (and unprepared for) the consequences of your actions, like Mary Shelley's Frankenstein. There is a long list of "good idea" species introduced intentionally into ecosystems throughout North America, most of whom are still causing problems: kudzu, feral hogs in the South, starlings, nutria, mongoose, monster carp, Siberian elm, tamarisk, Russian olive...

John Muir once wrote, "When we try to pick out anything by itself, we find it hitched to everything else in the Universe." Scientist and author Barry Commoner called it the first law of ecology: everything is connected to everything else. Human attempts to manage (or introduce) just one thing in nature often cascades like dominoes.

A less human-centered philosophy for caretaking nature is Aldo Leopold's "land ethic": "A thing is right when it tends to preserve the integrity, stability, and beauty of the biotic community. It is wrong when it tends otherwise." Leopold importantly added moral and ethical obligations for humans' treatment of nature. He also recognized that natural elements have a telos, or an end of their own, an autonomy in their ability to self-direct and -regulate. In other words, humans are not needed to *manage* natural communities—and should at times not step in and even try. He described this shift as being able to think of a muskrat as a subject and a being with its own purposes, activities, and perspectives, not as an object of human interest or gain.

Leopold's perspective is quite an improvement to typical *natural resource management*, though it still regards humans as the arbiters of right-versus-wrong actions and values. If humans are the only true value-assigning entity, then before humans arrived on the planet, there was nothing of value here. It's like claiming there can be no history without human historians, or no biology without biologists. As philosopher Holmes Rolston noted, the scales of value we construct do not constitute *value* any more than the scientific scales we devise (inches, gallons) create what they thereby measure. It is very hazardous in a time of ecological crisis to say that one species chooses itself as the absolute valuer of everything else in nature (cattle more valuable than wolves, ladybugs more than mosquitoes). It's a bit like holding a mirror and seeing your own reflection in the center because you're the one holding the mirror.

My rewrite of the bumper sticker: *Wolves: predators, just like you. Can't control 'em, so enjoy 'em. Everybody's gotta make a living, everyone plays a part—just like you.*

The Swedish language has no word for *land management* so it uses the Swedish word for *caretaking*. So I would rename my former Natural Resource Manager job title as Caretaker-Friend of Earth Companions. Sounds like a foreign language, indeed.

Geo-engineering

"Geoengineering is the deliberate large-scale intervention
in the Earth's natural systems to counteract climate change . . .
Generally, they can be grouped into two categories: Solar
Radiation Management or Solar Geoengineering, and
Carbon Dioxide Removal or Carbon Geoengineering."

—Oxford Geoengineering Programme website

"Managing sunlight" seems the stuff of an outlandish James Bond movie where he battles an evil megalomaniac out to destroy the Earth. But it is not. While pockets of people dig in their heels and pretend climate change is not real, prestigious Oxford University is gambling its scientific reputation on engineering the planet out of it. I landed on the Oxford website while searching for assessments of the potential dangers of geo-engineering.

Every time I read a geo-engineering scheme, I shake my head: launching trillions of little sunshades into orbit to reflect sunlight, ocean vessels spurting seawater into clouds in the hopes that they reflect more sunlight, dumping massive amounts of iron into oceans to spawn carbon uptake. Instead of focusing on the cause (humans' combustion of fossil fuels), geo-engineering focuses on the symptoms (too much heat and too much carbon dioxide), even though economists say a focus on the cause would be far, far cheaper.

In my searching, some scientists extolled the dangers of geo-engineering (and its potential to unleash innumerable *new* global problems), though more scientists seemed to defend its necessity. The pro-geo-engineering reasoning followed several veins: it's an exciting new frontier and a new puzzle to solve; it's the changed Anthropocene so let's change the Earth more; and emission-reduction ain't happening so we gotta do this. One engineering and ethics professor called geo-engineering a manifesto, a duty to "manage and reconstruct" Earth's "human, natural, and built systems."

Geo-engineering may well be the mother-of-all-metaphors for controlling and managing nature: engineering the amount of sunlight reaching the earth, and engineering carbon, the common element of every being, living and dead. Geo-engineering is superficially attractive because it delegates responsibility to engineers—no need to change our behavior and energy-use policies.

These schemes sound audacious in part because they target things that seem so distant to our lives: the planet and the upper atmosphere. The scientific language is passive voice, with emotion and living beings removed; it speaks of "solar radiation," "emissions reductions," and "Earth systems engineering and management." Yet, when you recognize that humans are embedded in nature, geo-engineering is not distant but right in the backyard.

One geo-engineering scheme to "manage solar radiation" would inject the stratosphere with aerosols. Even if sulfur aerosols did deflect sunlight, they would eat away at the ozone hole and increase acid deposition on land and oceans, which includes where you live. Aerosols would also whiten the sky above your house; one scientist admitted that the disappearance of blue skies "could have psychological impacts on humanity." The sun shields way up there might seem as harmless as putting up shade umbrellas. If they actually worked, decreased sunshine would mean—just in human terms—less sun for your geraniums and tomatoes and less sun for your neighbor's solar panels.

These schemes have given me nightmares in daylight and darkness; I wonder whether my dad the scientist would ever have conceived of a time when science overreached, when it encountered lines over which it should not step. Tampering with sun and ocean seem to be examples of Nietzsche's "tragic insight," when he glimpsed science reaching some limiting point when scientific knowledge would recoil upon itself and bite its own tail.

The lexicon surrounding environmental problems often frames them as somehow the fault of "Nature," which as Larson says, means that science becomes both the savior and the solution, as if the problem lies "out there" and not "in here" in ourselves, in our social world, and in how we relate to nature. The "deliberate large-scale intervention" on Oxford's website suggests that engineers must rescue the planet's natural systems, even though science and technology (and population growth) contributed mightily to their demise.

Bound up in geo-engineering language are two key metaphors, *progress* and *rationalism*. We view *progress* as a linear process where future change is always positive. You might think of progress as curing cancer, earning more money, growing a business, or creating a "better life" for your kids, never envisioning potential negatives en route. *Rationalism* holds that reason and science are what create knowledge. You encounter rationalism if you are told that emotion and wisdom are not proper ways to "know" something. These metaphors stem in part from beliefs of eighteenth-century Enlightenment and the Age of Reason: human beings did (or could) possess knowledge that would enable them to remake and master the world. Political theorist Bob Tostevin said that once science uncovered scattered pieces of what makes nature tick, it was easy to forget nature's essential independence from human will, and easy to conflate the control of nature with science— even if just our *faith* in science.

In *The Future of Everything: The Science of Prediction,* Canadian writer and mathematician David Orrell wrote,

> We in the industrialized world still tend to see the world in objective terms, as something to be manipulated and controlled, slave to the laws of cause and effect. This is the shadow side of our great inheritance from two millennia of science . . . By turning the world into an object we can control, however, we also deny it life. . . . We will choose to protect nature only if we value it—and not just as an object, but because it is alive. The only way we will respect it is if we understand that we cannot control it.

It is hard to value nature with a language that delineates science "facts" that simply *are* as distinct from "values" that *ought* to be. "Objective" science seems to refer to external things, whereas "values" reside internally. But the boundary is deceiving: sometimes science metaphors seamlessly incorporate values, such as "health" of a watershed, a desirable cultural "ought."

A refreshing way to marry the facts of science and the values of nature comes from "indigenous science." The Little River Band of Ottawa Indians came together to release lake sturgeon back into the Great Lakes, a tribal ceremony that was integrated with scientific reintroduction. Indigenous science begins with a recognition of collective values (such as how to approach and honor the living world) rather than an external definition of nature's "problems." "Resources" become relatives and subjects, and humans (including scientists) are respectful partners or siblings in interconnected communities. "Facts" derived from ecosystem science are just one element of cultivating moral relationships with Kin.

This is a new approach for science, but not for life. For any major life decision, I gather "facts" but also rely on my intuition and the feelings of others. Although many see emotion as alien and contrary to science, research in neuroscience, psychology, and risk communication has found that emotions are key to moral decision-making, whether about geo-engineering or nuclear waste. Emotions convey important information regarding risk perceptions that *rationalism* alone cannot. Sabine Roeser, a professor of ethics and philosophy, concludes that emotions are a necessary source of reflection and insight concerning the moral impact of climate change, not a threat to rational deliberation.

Emotional engagement, she concludes, leads to a higher degree of motivation and urgency than a detached, rational stance on climate change.

My rewrite of "geo-engineering" is aimed not at Oxford engineers, but at those of us who must evaluate the language and substance of these schemes:

The problem is not our home the Earth but our relationship with her. Science has told us that our prodigious use of fossil fuels has given her a temperature, which has disrupted her climate cycles and the ecosystems that support the existence of all beings. What we need to fix is how we value her as an endless storehouse of personal materials, instead of as a precious living being who grants us precious elements—and joy—in our lives. This has created a hole in the fabric of our community, the community of all living things. The "fix" cannot be delivered by science and "geo-engineering" but must come from our hearts and minds.

A Language of Connection

"Better living through better chemistry" was a popular TV ad slogan of my youth. It was an apt articulation for a society in love with gadgets and consumer goods created by science and technology. As a daughter of a chemistry professor and who later worked in "natural resource management," I am a product of the anthropocentrism that viewed the earth as a "resource" and of a language that distanced and neutered living things as *its*.

My rewrites attempt to model *ecocentrism*, which sees all entities (living and nonliving) as intrinsically valuable. Humans are an interdependent, integral part of the biological world but are not superior nor dominant over it. This worldview lies galaxies from current Western beliefs, but strategies for a language (and practice) of connection are a good place to start.

First, all languages must find ways—through metaphors, pronouns, structure, and vocabulary—to be languages of connection. There must be space in the language (and in our hearts) to expand the boundaries of caring to land, water, creatures, and atmosphere and grant them moral consideration. Author Scott Russell Sanders called it a "fellow feeling" of kinship and kindness to fellow beings. And Kimmerer extends that feeling to place: "To be native to a place we must learn to speak its language."

Kimmerer suggests some universal pronouns for fellow beings. She notes that there is already a perfect English word for plural beings: *kin*. For the singular form, she coined *ki* from the Potawatomie word *Aakibmaadiziiwin*: a being of the earth.

Some years ago, I began to refer to all animate creatures as *who* instead of the "grammatically correct" *that*. I also attach *he* or *she* to individuals—accurately when I can, but if I don't know whether the snake is male or female, I alternate genders. (Alternating genders is what I have long done for humans instead of the universal default to *he*.) Thus far, only one editor has given me any grief. Most people do not seem to notice the words in my speech, although I hope my choice is affecting.

A second step is the need for gratitude and love. In *Love Letter to the Earth*, the Buddhist monk and teacher Thich Nhat Hanh articulates the empathic leap of a large, encompassing, loving relationship to the entire planet:

We often forget that the planet we are living on has given us all the elements that make up our bodies. The water in our flesh, our bones, and all the microscopic cells inside our bodies all come from the Earth and are part of the Earth. . . . Knowing this, we can begin to transform our relationship to the Earth. We can begin to walk differently and to care for her differently. We will fall completely in love with the Earth. When we are in love with someone or something, there is no separation between ourselves and the person or thing we love. We do whatever we can for them and this brings us great joy and nourishment. That is the relationship each of us can have with the Earth.

A similar love, *biophilia*, is the love of all that is alive and vital. Biologist E. O. Wilson claims that biophilia has ancient roots as the instinctive bond between humans and living systems, and humans subconsciously seek connections with the rest of life. At an immemorial, core level, humans' natural love for life helps sustain all life.

For Kimmerer, empathy and love call for gratitude and giving back. She recalls during canoe trips her father pouring the first cup of coffee onto the earth as thanks for all the gifts received. And she describes the Potawatomi tradition of *minidewak*, a giveaway ceremony that is a favorite at powwows. The honored one does not receive gifts but is the giver of them, demonstrating wealth not through hoarding but through giving away. Reciprocity is a reminder that all flourishing is mutual. It is a moral covenant that "calls us to honor all our responsibilities for all we have been given, for all that we have taken. . . in return for the privilege of breath."

Finally, we need to see ourselves as embedded and embodied in nature. John Muir once wrote, "Most people are on the world, not in it—have no conscious sympathy or relationship to anything about them. . . touching but separate." It's a fitting image for many of us modern humans: brushing the surface of the earth and feeling largely separate from it.

"The relation of self to setting," said zoologist and philosopher Neil Evernden, is the only relevant relationship in a discussion of humans and nature. This is not a causal connection but fundamental interrelatedness that recognizes that discrete entities and boundaries are illusory. To practice that in everyday life is to see that we all "infect" one another, human to earthworm, that we continuously inhale, absorb, and exhale the stuff of the earth through porous boundaries.

Another avenue is to change perspective, to see the Subject in other beings. To see what the muskrat sees, Aldo Leopold buried himself—prone—in the muck of a muskrat house just to have the muskrat's muddy vantage-point of the marsh around him. With his eye at pond level in the muskrat's house, he saw what the muskrat saw, he smelled what the muskrat smelled, perhaps he imagined what the muskrat would do next. Leopold acknowledged that without interference or guidance, the muskrat had the knowledge and ability to sustain itself, build houses of marsh muck, and live its muskrat life. With the cold spring mud soaking into his clothes, Leopold no doubt knew that the builder of that den was the only creature who knew how to be a muskrat, and left to its own devices, would be precisely that.

Charles Foster chose a badger den for a changed perspective. This UK veterinarian, with a doctorate in medical law and

⊙ ⊙ ⊙

⊙ ⊙ ⊙

⊙ **8** ⊙

Implosion at the Mall

"**N**eed some Jesus with your jeggings?" quipped the *Huffington Post*. Then head to the newly opened City Creek Center, an upscale mall in downtown Salt Lake City, built and owned by the Mormon Church. The mall has Tiffany diamonds and Porsche watches, Vegas fountains of fire and water, and views of the Mormon Temple. And it has a "1,200-foot authentic re-creation of historic City Creek"—complete with two hundred trout—flowing between rows of shops.

Much of the flurry of media coverage for the mall's grand opening noted the saintly virtue bestowed on shopping here. The *Huffington Post* said the place gave off a biblical air with fountains that shoot water and bursts of flame. The *International Business Times* said, "New rules, same old retail at Utah's 'Mormon Mall'" (referring to restricted alcohol sales and being closed on Sunday). And most every story mentioned the mall's "re-creation" of the creek so important to Brigham Young's followers long ago.

I generally shun malls, but I had to witness the implosion of religion, shopping, and "nature" in this one. The lack of cultural boundaries between entertainment, consumerism, and faith was fascinating. I needed to understand why the city's namesake creek, the original lifeblood of Mormon settlers, was now a spectacle in this temple of shopping. City Creek Canyon rises behind the city through shrubby foothills to moist pine ridges, but its water was banished a century ago into the storm drain before it reached its city. The water flowing through the mall is treated city water, not creek water. How did City Creek—this piece of everyday nature that people (and other animals) loved and valued—arrive at the mall?

~oggoge

Within a day of arriving in the Salt Lake valley in July 1847, a small advance party of Mormon pioneers had dammed the south fork of (what they eventually called) City Creek to irrigate rows of just-planted potatoes. Within the month, the Saints built another dam to form an outdoor font for rebaptism, where newly arrived emigrants could immerse their bodies and redeclare their personal faith.

One hundred and seventy years later, just a few blocks from where the first shovelfuls of "fertile, friable loam" were dug, stands City Creek Center. The mall brochure says it's "a place where a sunlit creek meanders through fountains, over waterfalls, and alongside ninety of your favorite stores and restaurants. A new shopping experience where a retractable roof above an open-air promenade creates the perfect climate in any season. A place where urban elegance meets outdoor grandeur. More than a shopping center, this is a city center."

The Salt Lake valley is a self-contained watershed at the edge of the Great Basin without a hydrological outlet to the sea. The towering Wasatch Mountains, whose snowy peaks were the backdrop for the 2002 Winter Olympics, rim the east side. Prodigious snowmelt runs down Wasatch canyons in seven major creeks, each flowing through the city (now, largely underground in pipes) to the Great Salt Lake, their final destination.

Mormons, fleeing from the Midwest, were looking for their Zion, a place to transform into Eden. In Salt Lake, they found an oasis at the edge of semidesert; they knew that its waters were a Godsend. The church's second leader, Brigham Young, declared (though no one heard the utterance or could confirm it), "this is the right place" to settle.

In 1847, the land was "owned" by Mexico, though it was the longtime home of large tribes of American Indians who knew the fertility of the oasis zones along the mountains, which had sustained them for centuries. Ute Indians knew City Creek as

No-po-pah, or Elk water. Yet according to an Epistle from Brigham Young two months after they arrived, Salt Lake valley land was to be given to the Saints as an "inheritance," though they lacked any legal title. A later church document granted Mormons the exclusive use of City Creek.

⌇⌇

On my first visit shortly after the spring grand opening, I drive into the gullet of underground parking; the garage holds five thousand cars, which makes me feel somewhat queasy and ensnared. The escalator deposits me street level near an immense bronze sculpture, a tall mound of creatures positioned above brown metal ribbons of water. Two seagulls stand stoically in front of a tall beehive. Behind them, a bear emerges above trout lying on their sides on a bulge of water, their mouths agape. Behind them, three river otters swim belly-up on the water. On top is an eagle with its wings spread forward. "Stream of Life" depicts the "relationship of local wildlife with creek-side habitat."

The beehive is a curious, but predictable, inclusion. Utah is the Beehive State, and the beehive adorns the state seal and flag, buildings, manhole covers, and a fair number of businesses. An 1881 article in the *Deseret News* explained the symbolism: "The hive and honey bees form our communal coat of arms. . . . It is a significant representation of the industry, harmony, order and frugality of the people, and of the sweet results of their toil, union and intelligent cooperation."

On a bench surrounding the sculpture, I thumb through a supplement to the *Salt Lake Tribune* about the spring LDS General Conference (LDS from the formal church name, Church of Jesus Christ of Latter-day Saints). Reporters profile Mitt Romney as an LDS bishop and stake president, they report "How Utah's Legislature Marches to Mormon Beat," and they speculate that the mall would be a new LDS conference hangout: "This weekend, the faithful will follow a creek—a City Creek. And what a revelation it is. . .[After years of construction, Mormons] won't recognize their Zion. . . . In between testimonials, conference attendees will find near-sinful temptations in every direction—between Tiffany diamonds, Porsche watches, Gucci boutiques, Godiva chocolates, and a Cheesecake Factory."

Though the entire church complex—tabernacle, temple, HQ, and gigantic conference center—is right across the street, the mall signage and PR materials don't mention the church or its ownership of the mall. The LDS Church invested about $1.5 billion (and its developer spent another $75 million) to build the mall complex (which spans two city blocks and includes 600 condos and apartments and 1.7 million square feet of office space). Press material made clear that church money came from City Creek Reserve Inc. (a church-owned private real estate company), not from the 10 percent tithing good Mormons pay. If Saints shop there, however, the mall's operating entity expects a 12 percent return on investment to the church.

<hr />

By their first spring in 1848, Mormon settlers replicated their utilization of City Creek on all seven Wasatch streams: an ox team plowed a channel for water to flow to fields, row crops, and dwellings. Creeks also powered gristmills and sawmills.

In a report to the U.S. government in 1850, Captain Stansbury described the City Creek waterworks: "Through the city itself flows an unfailing stream of pure, sweet water, which, by an ingenious mode of irrigation is made to traverse each side of every street, whence it is led into every garden spot, spreading life, verdure, and beauty over what was heretofore a barren waste."

Like settlers across the West, the Mormons were intent on replicating the moister Euro-American environs from which they came. It's a common human-centered focus, pursued with little knowledge of the environment (and its limits) and accompanied by the dogged belief that with enough sweat and engineering, anything is possible.

Mormons added religious weight to the charge: the Earth had been cursed with thorns and thistles, and their task was to redeem it by changing what they saw as a barren environment into a productive, green Eden. As one scholar said, "For the Mormon Church, it was a religious obligation to change the landscape;

settling, building, and producing were religious duties and not simply a means to subsist."

Historical (as well as modern) writings about Mormon settlement like to invoke Isaiah 35:1: "The wilderness and the solitary place shall be glad for them; and the desert shall rejoice, and blossom as the rose." What's curious in the popular retelling of making the desert "bloom like a rose" is that the Salt Lake valley was never a desert but a fertile, semiarid oasis with springs and snowmelt flowing into the valley year-round. The rich, dark humus of millisoils grew lush pastures of perennial bunch grasses before the sod was cut for irrigated crops. Yet by the late 1800s, Utahns had reconceptualized their richly endowed land as if it were the great American desert in need of salvation.

By 1876, water tanks and cast iron pipes delivered City Creek water directly to household taps. The *Deseret Evening News* proclaimed, "Pure water running into your own house. . . . The fall of the stream is rapid, and sufficient, without artificial pressure, to take it into the highest rooms in the main portion of the city." The creek flow varied by season but was generous; an engineer estimated ten million gallons a day.

The development of such modern convenience represents a move from an Aquarian Age to a Hydraulic one, as historian Donald Worster labeled it. When water moves underground and indoors, he concluded, our knowledge and intimacy with it move underground as well. Without the everyday, visual, visceral connection—of snow and rain and creek—most people only know that potable water arrives via one set of pipes and waste leaves through another.

~ﾟﾟﾟﾟﾟﾟ

I walk "downstream" into the mall and under the end of the closed glass roof.

"Bryce! I found a fish!" a young girl yells to her brother.

Just past Kay Jewelers, the creek picks up speed, and a short stub veers off the main creek to stretch around the visitor information booth. The water is deeper here, and a dozen large trout swim in lazy loops. Youngsters perch on sandstone boulders lining the creek, watching and pointing. A young man makes a sweeping motion with his forearm as if fly fishing, and his girlfriend smiles. An older woman said to her husband, "Well, will ya look at that."

The simulated creek "flows" via six invisible pumps that circulate the tap water; for the section with trout, the pH is adjusted and the water heated. The stream-bottom rocks here have algae on them, unlike the chlorinated sections of creek. The mall website said all the plants are "from" the creek, but the only one I have seen growing in City Creek Canyon is watercress. There is no dirt on the stream bottom or banks, just tidy stones and pebbles that don't tumble with the current.

"Bryce, come here! He jumped!" the young girl, now standing on a boulder, yells.

Along the walkways between creek and stores are dark bronze trash and recycling cans with cutouts of reeds on their sides. The same "reeds" form the bronze railings on the level above. Sandblasted mule deer tracks walk stream-side, stopping abruptly at a big boulder with "No Climbing" sandblasted on one side, where a young couple sits.

From the backside of the information booth, water spills out between the thin sandstone bricks, like information leaking into the creek below. At one end of the booth on a tall pole are two electronic screens—the bottom, a scrolling list of airline flight departures, and the top, a weather map of the United States. I walk closer. A dark blue mass inches jerkily over central Idaho, looping back then jerking forward; precipitation is falling somewhere. Salt Lake temperature: 58. No delayed flights. The weather map—indeed weather itself—seems superfluous here. The retractable glass roof covers half the block (and "reduces the demand for heating and air conditioning"), yet every store door was propped open, wafting heated air.

At the far west end under the glass roof are clusters of small aspen trees. In this seasonless place, snow will never collect around their trunks, and their leaves will never quake in the breeze. Will they know when to burst their buds or turn their leaves to butterscotch? The transplanted nature here—creek, light, plants, precipitation, trout, trees—cannot behave as nature intended. Weather and seasons are replaced by monitors, put under glass, controlled by water pumps, adjusted by heaters. This is nature as a design template, like Early American or Art Deco. Yet the message is that the mall's nature imitations and re-creations appropriately honor the original "Stream of Life."

From a creek-side bench, I watch kids in strollers and people with shopping bags, some crossing the creek via footbridges. It's pleasant along the water, and it's very bright even with the glass roof closed. It's more peaceful than a city sidewalk, but it does not remind me in the slightest of the wilder canyon "upstream."

The upper reaches of City Creek Canyon are dense with white firs and Douglas firs, while Gambel oak and cottonwoods dominate farther down. I have seen mule deer, elk, porcupine, grouse, lots of birds, and coyote up there; friends once saw a cougar sleeping in the crook of a tree. Slowly and steadily over twelve miles, the canyon climbs five thousand feet. Past the water tanks and water treatment plant, the canyon closes in and cools, and the city feels very far away. It's not wilderness; it's been used and overused, but it's a haven from city hubbub and noise. It provides authentic enchantment and bits of unpredictable magic like cougars in trees.

Mostly, the mall is just very noisy. It's a high-decibel background that sounds cannot pop through—except for the constant Muzak that is piped from (I finally discovered) the old-fashioned street lamps. I cannot hear the waterfall or people talking unless I'm very close.

Just like every other mall, this one has a Gap, Limited, Footlocker, Disney, and anchors like Macy's and Nordstrom. Only a few stores here have Utah origins, one being Deseret Book, the Mormon bookstore. The creek is the enchanting and unique feature in this mall, and the water and shouts about trout draw people near.

The press packet touts the sidewalk snowmelt system, a four-sided outdoor gas fireplace that provided heat and ambience, and the glass roof. Nevertheless, it boasts of the mall's green LEED certification and "sustainable design."

The mall PR said the glass roof was the only one outside of Dubai, but glass-covered shopping areas are centuries old; one in Paris was called part vaudeville and part religion. In the early 1800s, shopping in Paris was difficult because the narrow streets lacked sidewalks. So between 1800 and 1830, seventeen Parisian arcades were created: streets were closed to traffic and covered with glass awnings. Narrow, marble-walled passages within and between blocks connected large areas of shops. The arcades were called consumer "dream worlds" with goods, food, drink, vaudeville entertainment, gambling, and prostitution.

In the 1920s, social theorist Walter Benjamin called the Paris arcade "the original temple of commodity capitalism." Benjamin compared the arcades to religious structures; he described one as a "nave with a side chapel," and he called department stores temples consecrated to religious intoxication. The wide array of goods evoked enchanted dreams and wishes; at the same time, they functioned as nightmares. He called it *phantasmagoria*—an exciting cornucopia of one's wildest fantasies, yet full of deceptive illusions and specters, a place where profusion seemed within grasp yet was impossible to reach.

A fat trout breaks the water in front of me, investigating a green bit floating on the surface, then darts under the footbridge and out of view.

Sociologist George Ritzer also linked shopping and religion. In *Enchanting a Disenchanted World,* he said that "cathedrals of consumption" like modern shopping malls have a lot in common with religious centers: fulfilling people's need to connect with each other and with God's creation. As highly efficient, standardized, and interchangeable places, malls (and to an extent, religion) must work to maintain enchantment in the face of increasing rationalization and disenchantment. Simulating nature is one way to do that. A recreated creek is a controllable and dramatic public display and can be made more predictably spectacular and enchanting than its authentic counterpart—like monster trout visible below the live weather map.

I walk along the creek past a tea shop (where fox prints cross the walkway) and take the escalator up to the pedestrian skybridge, which spans Main Street and leads to the other full city block of City Creek Center. The skybridge is noisy and hot, and voices echo off the low-iron glass etched with leaves: the view beyond is the gray spires of the LDS Temple. On the skybridge floor is a brass map of Salt Lake City in 1871, its neat squares laid out in a grid stretching up to the curled, calico pattern of foothills. In the western portion is an undulating, unmarked ribbon: City Creek. A plaque on the wall says the city survey was designed to follow Joseph Smith's plat of the City of Zion, and that Brigham Young walked between the south and west forks of City Creek and pronounced that a temple would be built there. What's not mentioned on the sign is the religion these men practiced, nor the fact that the temple plans meant that the south fork needed to be moved out of the way.

Below, an electric TRAX train pulls away from Macy's, which wears the original facade of ZCMI, the letters imprinted above tall iron columns and gold filigree with its founding date, 1868. Zion's Cooperative Mercantile Institution was considered America's first department store.

After twenty years in the Salt Lake valley, Brigham Young knew that the Saints' geographic isolation was ending and he sought a way to unify their economic lives. From the pulpit, he told the Saints not to buy from Gentiles (as non-Mormons were called) who had settled in the valley: "Cease to buy from them the gewgaws and frivolous things they bring here to sell to us for our money and means." Young proposed a Mormon mercantile system that would require cooperative effort and reap cooperative benefit. Individual Saints were encouraged to buy small shares in "the People's Store." He believed that if profits were not distributed among the people, it would divide the community into classes and produce "the hateful and unhappy distinctions which the possession [or] lack of wealth give rise to."

By 1873, ZCMI sales were an astounding $4.5 million a year. By the end of the decade, over half a million dollars was declared in dividends on an original investment of $280,000. ZCMI co-ops (supplied by ZCMI Wholesale) sprang up in Utah and neighboring states under the logo of Zion's Cooperative Mercantile Institution: "Holiness to the Lord" arched above an all-seeing eye. Until 1895, the church controlled and promoted its co-op department stores. In the following century, the church maintained virtual control through stock ownership by prominent church members or by the church itself. In 1999, the LDS Church sold ZCMI to May Department Stores, which is now Macy's, and the old cooperative institution disappeared.

Although church teachings (past and present) stress frugality, thrift, modesty, and antimaterialism, the original ZCMI provided plenty of razzle-dazzle with three floors of glittering lights burning into evening hours (lit by its own electrical plant), an hydraulic elevator, and the first escalator in the western United States. Five horse-drawn wagons made deliveries in any weather. The newspaper advertised, "There are articles of use, and articles suggestive; articles simple, and articles recherché; articles for amusement and articles requiring intellect for their appreciation." As P. T. Barnum once noted, a great department store is much like a circus with its bright lights, colors, and flashy enticement. And like all spectacles, the bar must be progressively raised.

I left the skybridge to search for my entrance to the parking garage.

⁓

By the late 1800s, City Creek was in very bad shape. Industry; power generation; and mills for lumber, flour, and silk had greatly degraded the pure creek. In the foothills above City Creek, severe overgrazing by livestock caused erosion, and heavy rains and snowmelt delivered a slurry of topsoil and animal waste downstream. Overgrazing damaged area towns and farms from the mid-1880s through the late 1930s.

Beginning in 1907, Salt Lake City purchased lands in City Creek Canyon to protect the watershed. But the other "solution" was to banish City Creek from its city. Residents decided that City Creek was a nuisance once it arrived in the city, flooding streets and sometimes buildings if not protected by sidewalks during spring melts. When flood water drained, animal waste and mosquitoes remained. In 1909, City Creek was put in a culvert that ran underneath North Temple street from the canyon mouth to the Jordan River.

I visit the mall again late Saturday afternoon on the first day of fall, yet a hot 86 degrees. I sit again next to the Stream of Life bronze sculpture. A Hispanic young man with tight jeans and spiky hair lies in front of the river otters, holding a baby in a yellow onesie while a young woman takes their picture.

The small, shallow creek behind me is languid with pond scum, though I smell chlorine from the eighteen-foot waterfall in front of the food court. A couple strolls past slowly; the woman points to the waterfall and says to her husband, "Oh honey, look at that. Ever wanted one of those?" The waterfall attracts a steady stream of children and adults who bend down to touch the water. A brass plaque proclaims "mountain lion" above several sand-blasted paw-prints, which disappear near the food court doors.

Outside the food court on the open plaza, two young women in red-and-white striped knee-highs and blue straw hats swing a long jump rope while a young boy skips. His mother takes pictures. Next, an older plump woman steps forward but isn't able to complete a jump, despite loud encouragement from the rope swingers. A growing crowd stands watching, captivated with the street theater. With each jumper, the swingers count up the jumps, ending with a loud chorus of "ohhhh" when the person misses. Two other uniformed women move through the crowd with flyers. One hands a flyer to the Hispanic couple with the baby; her T-shirt read "Promo Team for Cirque du Soleil." Our enchantment was a marketing stratagem.

I walk through the food court to find the source of excited noise: the dinosaur play area, "inspired by Utah's prehistoric past." The room is filled with hard-plastic, smiling, plump dinosaurs in shiny cartoon colors, and it is overrun with running, screaming, and climbing kids. Interspersed are a few soft, flat, tan blocks (perhaps representing fossil pieces). Tables lining the play area are stacked with paper and plastic takeout trash. Malls have long promoted "hybrid consumption," adding features like food, entertainment, and childcare and play areas to keep customers shopping longer.

Up the stairs by the food court waterfall is the four-sided fire-place, open to the sky. Several long couches rim the fireplace, full of teens slouched and absorbed in their electronic devices. The music here is piped from the fireplace instead of the streetlight. At the top of each bronze fireplace grate it says, "City Creek Established 2012." Above the grates is a granite mantle with a sandblasted poem, which begins: "I HAVE NEVER SEEN SUCH A LOVELY SIGHT AS THE CREEK THAT IS FLOWING WITH SPEEDY DELIGHT . . ."

I laugh aloud at the bad verse and two teens look up.

North past the top of the waterfall and directly across the street is a large gray building with "Church of Jesus Christ of Latter-day Saints" etched above tall Roman Ionic columns. When I stand next to the creek, flowing tightly here through sandstone boulders and tall grasses, it look like it emanates right from the building. Mountain goat tracks walk from the direction of the church building into the mall.

Where City Creek vanished into the storm drain was eventually called Memory Grove after the various war memorials built there. According to a history of City Creek, the park fell on hard times in the 1960s and '70s because of the "hippies, runaway youth, drug dealers, and other questionable individuals" who congregated there, which later included the "Gay Community."

The creek rebelled as well. On Saturday, May 28, 1983, an abrupt spring melt of an abundant mountain snowpack over-whelmed the storm drain in Memory Grove, and City Creek revisited its city, roaring down State Street. Churches dismissed their congregations that Sunday morning to help fill sandbags. For a week, trout slapped against walls of sandbags as the muddy torrent carried them past ZCMI, office buildings, restaurants, and pawn shops. The spectacle of floodwaters through downtown made national news.

An aftermath of the flood (in addition to bolstered engineering) was the cautious suggestion that perhaps City Creek should be "let out" more often. Perhaps the stream should permanently run down one side of State Street. Maybe, as the *Deseret*

News suggested, "the presence of water could be designed into the fabric of the city in a more meaningful way."

~ℓℓℓℓ

Weathered wood boards cover the front of Anthropologie, and inside is an eccentric mix of women's clothes, dishes, linens, off-beat magazines, and antique-looking furniture, such as painted old wood chairs for $980 each. I put my handbag on a long bench covered with burlap food bags as upholstery and take out my sweater (it is freezing inside); I wish I had earplugs (the hip world music is blasting). I pull out a few hidden price tags: a $160 sweater and a blouse marked down from $168 to $79.

In another shop, I tell the young blond clerk that I had to check out a store called True Religion Blue Jeans in a mall owned by the LDS Church.

She quickly replies, "Oh, it doesn't have anything to do with that." She says the Los Angeles store founder said, "there was only one real religion and that's people, and that all people wear jeans." She then launches into what made their jeans so special. "Each pair is made by a real person, not by a machine. And it takes like two weeks to make each pair, and they will never wear out. They really do last a whole lifetime."

"How much are they?"

"They start at around one-hundred-sixty dollars, and the most expensive pair is like three-hundred-something."

Many women carry H&M shopping bags, so I visit the store. It is crushingly loud, this time techno pop-rock. The store is jammed with clothing racks and stacked displays and young people trying to move fast around obstacles while carrying the store's open-hamper shopping bag. Near some bright, slinky sweaters I hear, "You could like totally dress that up or down, ya know." I paw through a few racks; the prices are cheap and the sizes small.

H&M is an example of "fast fashion," which changed a twice-yearly fashion calendar into a constantly changing one with clothes made with cheap synthetic materials and low-wage labor in China (and now India, Vietnam, and Bangladesh). Some call the trendy, poorly made ten-dollar blouses and sweaters "landfill fashion."

The salesman in Porsche Design asks if I'd like to see the cuff-links I am leaning over.

"Oh, no thanks," I say. "I didn't know men wore these anymore."

"Oh, they're really making a comeback," he says. "And I think they look real classy."

"Is that watch really fifteen thousand dollars?" I ask, pointing to a case with a tag that read 15595, no dollar sign.

"Yes," he says matter-of-factly.

"Why? What makes it worth that much?"

"It has two movements. So when you travel to different parts of the world, you can switch to the other movement and not have to reset your watch."

"And that's worth fifteen thousand dollars?"

"To some people it is."

I shake my head. "That's more than many people in this country make in an entire year."

He smiles. "Well, these items are appealing to Porsche drivers. I grew up around wealth so I felt comfortable with it—I never had it, but I was comfortable with it. And hey, the wealthy have problems just like the rest of us."

A few storefronts away is Tiffany's. Two women and three teen girls, all plainly dressed, stand in front of a small display window with diamond necklaces draped on blue velvet stands. They are laughing, and I guess their language is eastern European. One woman pulls a small camera from her purse and gestures to one teen. The teen steps over to the window, one hand sweeping out to display the jewels like Vanna White, the other hand curving palm-down under her chin, grinning. Flash. They all laugh, then move on.

I so want to know what they said. Was the aping for the camera a mocking of such phantasmagoria, or was it a longing, a hunger for items beyond reach? Immigrants who come to this valley, then and now, are introduced to American riches and dreams, amazed by the profusion and tempted by the choices in our consumers' paradise.

In *I Want That!*, Thomas Hine concluded that we now live our lives in the *buyosphere*, an all powerful, encompassing, ubiquitous

arena where we are consumers in a marketplace rather than cit-
izens in a civic space. We are so enmeshed in consumer culture
that it is virtually impossible to leave the buyosphere (even in
wilder nature), which increasingly serves as the master logic of
social relations. Every item of culture—love, nature, sex, music—
can be transformed into commodity and subordinated to market
logic with a price tag. The consumptive hook of this mall is the
use of faux nature to get people buy the "dead" transformed nature
in all the stores.

In the Bible all people are equal in the eyes of God, but
the bounty of U.S. consumer culture is not heaped equally on
buyers in the marketplace. The top 10 percent have 66 percent
of the wealth, and 90 percent of us share the rest. That essentially describes a banana republic, which most people know only
as a store.

꩜

In 1997, City Creek was resurrected from its subterranean pipe
for another mile. The city and the Mormon Church spent over
a million dollars to create a concrete creek channel from the
Memory Grove storm drain to the newly created City Creek
Park on the northeast corner of the LDS office building. Rocks
were glued into streambanks that would never shift. Sidewalks,
benches, grass, and trees bordered the curving channel. The
tracks of birds and small mammals were pressed into concrete
and labeled. A waterwheel churned peacefully, dumping its load
without a true purpose but evoking a sort of pioneer spirit. Without a park sign, you wouldn't recognize it as the same creek
whose unfettered waters upstream form a narrow, shrubby oasis
in the foothills. The unearthed creek still doesn't reach City
Creek Center.

꩜

In the Main Street block of the mall, kids of various ages play
in the "Engage" fountain, a flat disc that shoots streams of water
briefly skyward in seemingly random patterns and timing, which
the "players" try to dodge. Everyone standing on the disc looks

wet. Another fountain has watery heads that look like watery jellyfish, which entices people to poke in a finger to disrupt the flow. The mall fountains were created by Bellagio, designer of Las Vegas fountains.

"What time does the fire fountain go off?" I ask a security guard. He is young with a round face and round belly, on which he rests his folded arms. "Flutter" is advertised as "mesmerizing with dancing fire on sheets of water spilling out in the shape of bells, in a fusion of pyrotechnics and nature."

"Oh, that doesn't go off until 9 p.m., right when the mall closes."

"Well, darn!" I say. "It's pretty dark now so I thought it'd be earlier."

"I know, but it keeps people around longer, shopping and eating. At least you can see the water part without the flames in a few minutes—that happens every hour."

With a soundtrack of heavenly soprano voices, the show begins. Streams of water reach upward in fluted bell shapes and collapse on themselves. At the edges, streams of water form domed lily pads. The patterns reverse, combine, and intersperse with syncopated single shoots of water high in the air. The chorus of high harmonized voices without words crescendos, softens, then grows. People take pictures. It lasts several minutes; I am disappointed.

I walk back to the security guard. "Since I'm going to miss the fire, can you tell me how it works—why the water doesn't put the fire out?"

He grins. "What keeps on burning even when it's on top of water? Think of camping."

"What, like stove-gas?"

"Kerosene. Mix a bit of water with the kerosene and it floats on top of the water. If you poured some on a river and tossed a match on it, it'd burn just fine."

I walk back over the skybridge. It is 7:30 on a Saturday night and the place is packed. The music-emanating street lamps light. Water in the deep part of the creek is churning as trout dart and roll to get morsels that kids toss. I ask one boy where he bought the fish food (thinking there were dispensers), but he looks at me blankly. He throws another piece; it's a french fry.

In the late 1990s, city planners studied how to "daylight" City Creek. The daylighting plan would divert City Creek from its existing culvert to an old right-of-way, bounded by three Superfund sites with high arsenic and metals concentrations. A 2003 Aquatic Ecosystem Restoration report said the daylighted creek would meander 7,900 linear feet in a dirt-lined open channel about three feet deep, ten feet wide at the top, and two feet wide at the bottom. It would flow just west of what is now The Gateway, an "open-air" mall with ninety shops, restaurants, fountains, office centers, condominiums, a planetarium, a children's museum, and movie theaters. The daylighting plan—thus far—remains a pipe dream. I'm not sure how I feel about this; I applaud celebrating waterways, but this seems an insulting freedom for such a faithful creek.

Several years ago, a friend invited me to the annual LDS Christmas show at the new Conference Center next to Temple Square. Our seats were in front of one of several large jumbotrons. The place held 21,000 people. Several balconies curved around the main floor and stage, adorned with hundreds of flocked trees, giant bows, and thousands of white lights. Behind the massive organ pipes, deep colors glowed and morphed one to another as the Mormon Tabernacle Choir sang. It was a tightly choreographed evening of extravaganza and fine music, and the closest I have witnessed to what some call the Christotainment found in megachurches.

It also was an example of what postmodernists call *implosion,* or the growing inability to differentiate among things and places because they are imploding into one another and losing their distinctness, their boundaries. Like religion and entertainment via jumbotron. Like trout entertainment and shopping in a church-owned mall.

Some critics say that ubiquitous consumerism is "the religion" of the century. Consumerism has reached far beyond the mere provision of material goods to the fulfillment of a pseudospiritual life; we look to the marketplace to provide answers to our problems and a sense of meaning and identity, perhaps more often than we look to a church. When church and marketplace implode, neither prevails. There is simply no clear boundary between them.

A few paces past the trout, I walk into Godiva. On the brown wall behind the display case is a gold sculpture of a naked woman with flowing golden hair astride a golden horse with flowing mane and tail. I buy two chocolate raspberry truffles for five dollars and take my gold foil bag outside to a creek-side bench. The truffles are good—but not sinfully good.

Earlier in the week, I heard one of my students tell her classmate, "I went to City Creek yesterday." My first image was of her hiking the canyon's creek-side path; then I realized her destination was the mall. It was the ultimate co-optation, that the creek's very identity had been skinned and grafted onto an imposter, an Avatar impersonating a creek.

There is no "nature" in the mall because creek and trout and all the accompanying pieces of nature-as-design-template have imploded, burst inward and mingled with blue jeans and truffles. There is no boundary between where shopping ends and nature begins; both amuse, sell, and entertain. Both make money for a church. To revere creatures with a sculpture, or pay homage by sandblasting their tracks into sidewalks, is to worship a false god. Build a waterfall; buy a waterfall for your very own. Human re-creations are equal—even preferred—to the Creator's originals (isn't that blasphemous?). "Outdoor grandeur" under glass.

In its canyon northeast of town, City Creek continues its faithful journey from snowfields and springs down, down toward its city before it's captured in treatments plants and storm drains. If the Mormons revered City Creek when they arrived 170 years ago, it was not because it was a wet jewel in a dry valley that nurtured Indians and cottonwoods and trout for centuries. Like so much of nature seen through pioneer eyes, it was worshipped because of what it did for them, what it gave up and is still giving up. The settlers appropriated the land from other people and other creatures, and later residents appropriated the creek—its waters, its name, its identity—for their benefit. In the western United States, water is survival, I understand that, but shouldn't survival come with gratitude and stewardship, especially by people of faith?

Jesus said it was easier for a camel to pass through the eye of a needle than it was for a rich man to get into heaven. Yet, the church built this high-end temple. Perhaps this implosion began long ago; Brigham Young once warned of frivolous things, but the department store he created carried the motto "Holiness to the Lord." It's one thing to preach thrift and to be content with little; it's another to view gewgaws as the just desserts of beehive industriousness.

I fold my gold bag with its gold cord handles. A small brown woman approaches, sweeping trash from under the long bench. When she reaches me, I lift my legs and say thank you; she sweeps under my legs but won't look at me. In the plaza where the jump ropers had been, a lanky black man plays electric piano and sings "Desperado."

The creek-side crowd has not thinned. The pageant that surrounds this phantasm beckons them in with water and fire, tempts with playthings and enchantment, and promises that salvation is just across the street.

9

Brushing Them Aside

Thwap! The black object landed solidly on my chest. I sprang from my chair and brushed it away. It landed on the deck railing and surveyed its new location, waving its long antennae. Oh, a pine sawyer beetle. Despite their name, they do not fell or harm conifers, just eat the needles and bark of trees that have died—good cleaner-upper beetles and certainly no danger to me.

I was embarrassed by my fearful reaction; I like these guys and think they are charming. Still, there is something about a two-inch insect landing on your body that can evoke what seems like a deep, instinctual reaction: eew, get away, I don't want you near my body.

When I worked as a naturalist decades ago, no one asked me about insects and I felt free to ignore most of them. However, as I watched the pine sawyer beetle I wondered about his life beyond the deck, where the beetle babies were, whether my avian visitors ate them. Invertebrates are a crucial part of many bird diets, but I did not know which insects sustained my winged friends. My habit of brushing aside insects (figuratively and literally) meant missing out on the world's smallest large force in the animal kingdom. I called Don.

Don Feener, a friend and biology professor at the University of Utah who studies the ecology of insect communities (particularly ants), had just returned from fieldwork in Brazil. I asked him to give me an insect tour of my backyard. First, he said, we must catch them, so we set a date for the following week to place traps. In the meantime, I schooled myself about insects.

Renowned biologist E. O. Wilson called insects "the little things that run the world." More than one million species of invertebrates (the vast majority of whom are insects) have been identified thus far, representing about 80 percent of the world's biodiversity. (That dwarfs the 60,000 vertebrates, which includes 5,000 mammal species and 10,000 bird species.) And, biologists estimate between 3 to 10 million insect species have yet to be found and described.

With that many invertebrates out there, ordinary people have helped fill the gaps in knowledge by collecting data as "citizen scientists." Part of the intrigue for writer and teacher Sharman Apt Russell to become a citizen scientist was that "you could spend a week studying some obscure insect and you would then know more than anyone else on the planet." In her book *Diary of a Citizen Scientist: Chasing Tiger Beetles and Other New Ways of Engaging the World*, she chronicles two seasons she spent tracking, studying, and rearing Western red-bellied tiger beetles. "Our ignorance is profound," she writes; we have so much to discover.

Similarly, Don Feener said he could hike up a nearby canyon and probably discover a fly that no one else has ever identified. Or, in Costa Rica, if he trapped fifty species of insects, chances were good that half of them would be brand new species. This suggest job opportunities (or at least citizen science fame) in entomology.

Even though they are small, on a per-acre basis insects not only outnumber but *outweigh* all the other animals combined. A single acre of land might host many millions of individual insects (and hundreds or thousands of insect species). On a plain in East Africa, ecologists found that just two species of ants were about equal in weight per acre to the combined weight of the large grazing animals there, like zebras and wildebeests.

Aristotle described "insects" as creatures with insections on their bodies, which remains a good taxonomic place to begin. True insects' "insections" are a three-segmented body (head, thorax, abdomen), to which are attached three pairs of jointed legs and a pair of antennae, and which are covered by a "skeleton" worn on the outside as a protective shell. (Think of the exoskeleton as a hard cuticle that keeps the bug from drying out and allows quick motion.) Other than those requirements, there is a lot of variation among insects: bees, beetles, dragonflies, ants, butterflies, flies, and true insects (like cicadas and water striders).

Some bugs I thought were insects are technically not, like centipedes and spiders. But they and insects all belong to the phylum Arthropoda, where everyone has some sort of exoskeleton, more legs and antennae than we have, and segmented bodies (but not necessarily three segments). The most accurate word for all of them is arthropods or invertebrates, but I'm guilty of lapsing into the colloquial and calling them all *insects*.

Many schoolchildren (including me) remember putting a caterpillar in a jar with some leaves and a twig. The caterpillar munched and munched, and then one day spun itself into a chrysalis (or pupa). After waiting for what seemed like forever, a butterfly emerged. The portion of this metamorphosis we didn't see was the adult butterfly laying eggs, which hatched into that caterpillar (also called a larva). A great many insects undergo this complete transformation: egg to larvae to pupa to adult. Some (like grasshoppers) skip the middle part and emerge from an egg as a small version of an adult (called a nymph) and then molt several times, shedding the too-tight exoskeleton before reaching adult-size. But when I think *insect*, I envision mostly the adult stage, which is really just the end of a multistage cycle that usually lasts just a day, a week, or several months.

It makes sense that insects "run the world" because the world evolved around them. Insects and other invertebrates were here on earth about 400 million years before we were and occupy virtually every habitat on land and freshwater—polar ice caps, deserts, hot springs, jungles, rivers, and backyards. Look out across a meadow, a forest, even the produce aisle, and know that it would look completely different without invertebrates.

Cornell entomologist Thomas Eisner called insects the most versatile of evolutionary innovators: "Pick an insect at random, and chances are there is something about the way it feeds, or defends itself, or reproduces, that is unique." He said insects have succeeded in one major respect where humans have failed: "They are practitioners of sustainable development." Although they eat plants, they also pollinate them, which provides a secure future for both insects and plants.

At the end of an afternoon of invertebrate research, I retrieved my laundry from the backyard clothesline. As I opened the door, the breeze ferried in a large flying insect. The next morning, I found it dead on the windowsill; a crane fly, I guessed. Its long legs were delicate threads, its wings lacy and opaque. By the following morning, the legs were scattered and wings lay above a body now greatly reduced in size. Other insects had been cleaning up.

Without invertebrates, we would be awash in dead and decaying material. They decompose our dead and recycle nutrients in the waste that we and other species produce. Without them, animal dung, plants, feathers, and hair would pile up and not decompose. Fungi and bacteria also depend on invertebrates to do their work. Pollinating insects make the world's crop plants possible, but less heralded are the actions of invertebrates at the other end of life, moldering plants, and working and aerating soil so that life and growth is possible once more.

⌇⌇⌇

On a Monday morning in May, Don arrived with the equipment. The area he deemed most insect friendly was the back of my backyard. Along the wooden back fence were several large maples and shrubs—pyracantha, quince, honeysuckle, and a dense one in the southwest corner I didn't know. Leaf litter and mulch covered the crabgrass and bindweed the sprinklers did not reach. It wasn't a monoculture of grass, which was promising for invertebrates.

To capture flying insects, Don erected a Malaise trap, named for a French entomologist and not the unease that is hard to pinpoint (though the bugs may have felt this once they flew into the trap). Several pieces of black mesh hung down to the ground; on top of them stretched a mini rain-fly of sorts made of white mesh that Don pulled taut and pinned with stakes. The white mesh rain-fly rose at one end like the prow of a ship, forming a chamber; at the bottom of the chamber hung an open plastic bottle filled with alcohol.

"Remember the movie *A Bug's Life?*" he asked. "Fly to the light! Fly to the light! That's what the bugs will do here." When insects flew into the black mesh walls, they would divert up to the light— the white mesh—which eventually would funnel them into the empty chamber where they hovered. When they tired, they would fall into the liquid below.

As a graduate student, Don became interested in ants in Costa Rica. "The tropics impressed me, with all the plants and all the ants. Everywhere you looked were different kinds of ants. Army ants are a big predator and they coordinate these amazing swarm raids." He also studied the energy costs to leaf-cutter ants from marching single file carrying all those leaf bits through the jungle. Essentially, he devised an ant tread-mill in the lab: he smeared the smell of a fellow ant inside a tube. When he put an ant in the tube and turned it, the ant marched, thinking it was following a fellow ant. Then he mea-sured its energy use.

After he erected the Malaise trap, Don installed pitfall traps to capture walking invertebrates. "This works great for digging the holes," he said, holding up the same contraption I use to dig holes for tulip bulbs. In each of six holes he sunk a plastic cup, its lip level with the ground. While he dug, I asked Don about his beginning entomology class.

"Most students take it just to get some lab credit, and they don't really know what they're getting into. But they seem to like the hands-on lab work and say they never knew insects were so interesting."

A freshman who enrolled in the class told him, "Bugs creep me out, and I really don't want to touch them." "But," Don said, "in the first few weeks, he got over it. And he did great in the lab and on the exams and turned in a great insect collection. Later, he got an internship doing mosquito abatement for the county and that turned into a real job. In a couple of years, he decided to go to grad school, in"—Don grinned broadly and turned toward me—"entomology."

So why are people so creeped out by bugs, I asked. "It's the way insects have been sold to us," he said, "and the way we feel we have to battle them with such concerted efforts. When people get to know them and touch them, they like them; they're converted."

"You got some dish soap?" he asked.

〰️

I never took an entomology class, but I have had a variety of insect experiences. For a 4-H project in junior high, I caught and pinned a collection of dragonflies. Fascinated with the praying mantis in our yard, I sketched one onto a piece of burlap and embroidered it for a wall hanging. I fondly remember fireflies, June beetles buzzing and clinging to the screen door, and walking sticks—an insect so shaped like a twig that we would shriek when one moved, but then tried to get him to walk on our arms. But I also remember bulbous ticks on the dog, scads of mosquitoes, and ever-present Daddy Longlegs in my bedroom and bathroom. My childhood contact may have been exactly the contact Don mentioned: if you get close to them, you'll find them interesting, even the gross ticks.

Invertebrate expert, naturalist, and writer Robert Michael Pyle told me that children on the whole are fascinated by bugs, in part because the lingua franca of childhood includes catching things. So what changes the fascination toward "small-scale wildlife" (as he calls them) to antipathy? "Some children pick up the fear fairly early on from various cultural messages. You hear about pests, you hear Dad talking about ants getting inside, see a Raid commercial on TV, you hear something at school. And, a couple of centuries of messages from agriculture and the chemical industry portray humans as being in a war with insects. A kid learns that these bugs are not your friends after all, as they were in childhood."

According to one study, adult attitudes toward arthropods fit a bell-shaped curve: on one end, a small minority take pleasure and likes insects and spiders, and at the other end, a somewhat larger minority expresses indifference. But the great majority in the middle feel apprehension, fear, and outright phobia toward bugs. Higher anxiety exists among children and females. The researchers concluded: people's fear, antipathy, and aversion are a big challenge in conserving arthropods and convincing people that these organisms deserve moral consideration like larger-scale wildlife.

The late Stephen Kellert (and longtime professor at Yale University) conducted a large study of human likes and dislikes in the animal kingdom. Not surprisingly, butterflies and ladybugs ranked high on the liked species, but wasps, mosquitoes, and cockroaches were rock-bottom. Factors that contribute to this: We do not like that they are dangerous and compete with humans, they are small, their bodies have foreign textures and structures, they move differently, and they reproduce like crazy.

Everyone has been bugged by some bugs: spiders and cobwebs in the basement, mosquitoes on camping trips, aphids on roses, and for the truly unfortunate, bedbugs or cockroaches. When I lived in the Midwest, I waged a constant battle with fleas on the dog. Moths recently destroyed a nice wool rug, and last winter I found larvae in my arborio rice. It's easy to use personal experience with household "pests" to brand all invertebrates as guilty by association. Our houses are sacred spaces to which we allow some to enter, but bugs are generally uninvited and unwelcome. The insects that affect our lives when and where we do not want them don't know they are pests; it's not personal. And of all the millions of insect species, only 1.5 percent do us (or our crops) any harm. Stereotyping all bugs as "bad" is like labeling all humans as criminals when very few are. (Come to think of it, the U.S. crime rate, nonviolent and violent, is double the "crime rate" of bugs.)

⌇ᵒᵍᵍˡᵉˡ

I handed Don the dish soap and he mixed a squirt into a watering can. He filled the six sunken cups up to the lip. The soap in the water broke the surface tension so the walking bugs (like beetles and ants) couldn't float or climb out and would sink to the bottom.

"You remember Uncle Milton ant farms?" he asked.

"Yes! We got one for Christmas when I was a kid. It was really cool," I said.

"Well, two guys from Hurricane, Utah were the ones who mailed you the ants after you bought the farm. And what they sent you was a kind of seed harvester ant, *Pogonomyrmex californi-cus*, that's found in Utah." Don chuckled under his breath. "But the ants they were sending to all these little kids—they really sting! One stung my foot once and the next day, the lymph nodes in my groin still hurt. They analyzed the venom and found it was the most toxic thing known to man at the time."

"Bizarre," I said. "And, it reinforces the stereotype that bugs are dangerous, even ants."

"Actually, ants are very reluctant to sting. You have to really provoke or threaten them. And some birds and horned toads eat them, so they have some tolerance of it."

As he packed up his gear, I asked him why he studied ants.

"I really admire them. Ecologically, they fill very important roles. They aerate the soil like worms do, they increase nutrient concentrations, they disperse seeds, and they're also useful preda-tors. And, ants are everywhere, and in great numbers. Their social lives and colonies are fascinating—the community has strife, they resolve conflicts, they make compromises."

He wiped his hands on his jeans. "Do you know where the intelligence of the ant colony resides, who really controls the colony?" I shook my head. "The intelligence and control doesn't happen until it's *distributed*—it's an emergent property of individ-uals interacting in a social system, a community. Now that's real intelligence."

⁓⁓⁓⁓

In the three days until Don returned, I wandered back to check the traps. The opaque Malaise trap bottle slowly filled with dark dots of various sizes; from the pitfall traps I removed leaves and twigs that had fallen in, but I couldn't see the bottom. I took advantage of a warm May day to work on a backyard project. In my digging, I encountered quite a few creatures: a couple of bugs rolled into tight roley-poley balls, along with three differ-ent kinds of worms and several different kinds of ants. A mil-lipede sped out of the hole, its feathery legs propelling him as if on a cushion of air. A beautiful small beetle with an irides-cent turquoise shell crawled through the leaf litter. A pale white moth fluttered by. I had no idea what were they all doing in my soil, my yard.

A month ago on a warm spring evening I sat outside talking on the phone. My dog sat cocking her head down at the grass, occasionally poking her snout to the ground to sniff. She'd walk to another spot and repeat. I searched with a flashlight and could see nothing, not even holes. Yet I am sure she heard movement, some-where down deep—maybe earthworms moving in slowly warm-ing soils or ants setting up colonies. It reminded me of a Chey-enne Indian saying, "When we touch the Earth, it's not just dirt. It feels like there's something pushing back."

Both inside and out, I started to see more invertebrates: a fuzzy dark spider trapped in the bathtub, hundreds of insects swarm-ing the porch light, a bright blue long and skinny fly perched on the hummingbird feeder, and other small flies and bees winging about. An infinitesimal dot of an insect ran down a page in my book. Fruit flies hovered near the bananas.

One afternoon after reading about invertebrates for sev-eral hours, I caught myself scratching my skin. I laughed; in true psychosomatic fashion, my mind was having a conversation with my body about bugs without my knowing it. The itching was a conditioned response, hard to extinguish; at some deep level, bugs to me meant bites and itching.

Scholars debate where we get our emotional responses to bugs, particularly fear. Some hypothesize that it's part of our evolutionary biology, that we possess an innate fear of a few potentially dangerous insects that we then generalize to all. I read about experiments where babies of various mammal spe-cies (who have not yet had time to learn of any danger) showed aversion to some invertebrates even when a threat was not obvi-ous. Another theory is "disease avoidance," which we developed from ancient fears linking bugs and disease—an association that is true for only a tiny percent of invertebrate species, such as some mosquitoes and fleas.

During the Middle Ages, spiders were associated with the plague because of their proximity to the real culprits: fleas who jumped off rats and other animals and bit humans. Spiders are usually neither poisonous nor the agents of illness they were thought to be. Current-day media depictions reinforce insect pho-bias, from the movie *Arachnophobia* to one I still remember from a teenage slumber party, *"Them!"* with radioactive, dinosaur-sized ants wreaking death and destruction.

༺༻

Last week, two companies left flyers on my doorstep for "pest control." One offered quarterly pest control service for $89.95. In spring, they sprayed for emerging "Black Widow Spiders, Earwigs, and house spiders" that were searching for "new nesting places." In summer, it was ants. "Ants are the most difficult insect to con-trol . . . Your ant problems range from them marching their dirty little feet through your food, to colonizing your kitchen." (My mom used white vinegar to erase the ants' scent trail.) What hap-pened to ant admiration in fables and proverbs, where ants are models of industrious, hard work?

Although the biggest threat to invertebrate populations is habitat loss, Pyle said the next biggest threat is the use of biocides, a word coined by Rachel Carson in *Silent Spring* to emphasize that such chemicals harm all living things, not just their targets. A report by the U.S. Fish and Wildlife Service said homeowners use up to ten times more chemical pesticides per acre on their lawns than farmers use on crops. Pyle said that lawn chemical compa-nies present checklists of "bad" insects and plants to induce civic shame and guilt to use their services. The products are so toxic, he said, that it should be considered child abuse to let kids play on treated lawns. And as Carson presaged in the 1960s, some insects have adapted to chemicals and become "super bugs," resistant to ever increasing amounts of toxins.

Pyle told me about a particularly worrisome class of fairly new chemicals called neonicotinoids that persist in the environ-ment, have been implicated in bee die-offs, and are now widely used on yards and crops. They are systemic, meaning the chemi-cals are absorbed throughout the plant or tree and are even present in pollen and nectar; the entire plant is toxic to any creature who touches or eats it. Residue from these chemicals can linger in the soil for months. Neonicotinoids kill pollinators like bees either out-right, or by causing problems in flying, foraging, and navigation.

Last year I took a picture to the garden store of large sappy holes in my peach tree trunk; peach tree borers, they told me, and pointed out the chemicals to use. I followed the directions in mixing and spraying, and wore gloves and a mask. The bottle said nothing about a risk to bees.

The Xerces Society (an international organization devoted to invertebrate conservation, which Pyle founded) called for manu-facturers to suspend the use of neonicotinoids on insect-polli-nated crops and to include warning labels about hazards to bees. The society also noted that the manufacturer-recommended appli-cation rates for gardens, lawns, and ornamental trees were over thirty times higher than rates approved for agricultural crops. I thought about the birds in my peach tree, the dog rolling in the grass beneath it, me eating the peaches.

I returned to the garden store and read the back labels' fine print for six kinds of neonicotinoids: Imidacloprid, Clothianidin, Dinotefuran, Thiamethoxam, Acetamiprid, Thiacloprid. Every bottle I examined had one. But names on the bottle fronts were not chemicals but characters: Killer. Hunter. Bandit. Knockout. Marathon. Control. Maxide. Surrender. Safari. Bug B Gon. Venom. Scorpion. Imicide. Macho. Malice. Sepresto. Widow. Wrangler.

The fear and loathing in those names surpasses a fear borne by our evolutionary history or a fear of disease or actual harm. Instead of hatred and total annihilation, there should be inverte-brate love and admiration for their pollination of everything in the produce aisle, everything flowering in our yards, and all the corn, soybeans, rice, and wheat. Without invertebrates in the soil—particularly ants and earthworms—it would be less aerated and fertile. And, insects sustain fish, frogs, mammals, and birds. Cor-nell University's Ornithology Lab told me that the vast major-ity of the world's 10,000 bird species consume insects regularly.

~※~

Don arrived Wednesday after work and emptied the pitfall traps, pouring each cup's contents through a small aquarium net; mostly mud and a few insects remained. He turned the contents into a bowl and rinsed it with a squeeze bottle full of alcohol solution. We caught three kinds of spiders.

The most pronounced insect phobia is of spiders, and if anyone should be afraid of them, it should be me. One warm Thanksgiving, I noticed a small red welt on my inner thigh; by the next morning I was in an InstaCare on triple antibiotics. I was bitten by a hobo spider, probably hiding in the downed limbs I had moved that morning. Hobo spiders were introduced from Eurasia, and Pyle said that although folk legends say their venom kills flesh (necrosis), there are no confirmed cases. I couldn't walk for two days and had a red, swollen, football-sized mark. But I don't freak out over spiders, perhaps because my childhood in the woods was full of them, perhaps because I loved *Charlotte's Web*.

However, many phobias are not based on any actual bad experience—just the fear of potential danger. Traditional approaches that seek to rationally control the fear through behavior modification or psychology often do not work. But a novel experiment reported in the journal *Neural Plasticity* used a spiderless therapy to cure arachnophobia in almost all participants.

Researchers gave each severely arachnophobic participant a CD of images; 44 were neutral images and 132 were images of everyday objects with "spider-like" features: an office chair with splayed legs, a tripod with articulated legs, objects hanging from strings, curving columns on a cathedral, a sculpture that resembled a web. Each participant viewed the slides twice a day for six weeks. Many subjects later reported they thought they were in the placebo group because their images did not cause any discomfort. At the end of six months, 92 percent of the participants were classified as nonphobic and many of them could approach a live tarantula; six even opened the lid of the cage.

The researchers said this spiderless therapy was aimed at the subconscious, automatic responses produced by the emotion centers of the brain. Real spider images would rapidly activate the remembered fearful response, but images containing "spider code" could gently produce plastic changes in the neural circuit. Fearful neurons were not activated, and the fear memory was essentially replaced with a neutral memory. Perhaps images of other distinct "insect codes" could help supplant insect fears with less fearful sentiments.

~*oggose*~

With a pair of fine tweezers, Don grabbed a wriggling creature from the mud of a pitfall trap—a sow bug, the rolled-up ball I saw earlier when digging. He explained how this invertebrate was related to the other creatures we would find.

"Okay, all these guys are invertebrates in the phylum Arthropoda. In terms of numbers of individuals, there are more ants than anything else. In terms of species, beetles are the most diverse. In the Arthropods are four main groups: hexapods (insects, including beetles), myriapods (centipedes and millipedes), chelicerates (includes spiders and ticks), and crustaceans, like this little guy. He's got gills and could exchange gases in the soapy water."

"So they like moist environments, I take it?"

"Yep."

In the muddy soup, a bright red dot stood out. "That's a mite," he said, "advertising its toxicity with that color. And here's an earwig."

"Gross," I said. Too many of these fellows had scuttled from the moist folds of newly picked garden produce across my kitchen counter and sink. And the pair of pincers was menacing. They really bugged me.

"Aw come on," he said. "They show great parental care. Female earwigs take care of their eggs until they hatch into nymphs, which is rare for insects. And if the young don't then leave the nest, she eats them. That's what I call tough love." I looked up from the bowl; Don was grinning.

He poked through the muddy collection: in addition to three kinds of spiders (which we needed a microscope to fully identify) was a millipede, a black beetle, and three kinds of ants, including a pavement ant. (At my last house, tiny pavement ants would swarm up from below the sidewalk, though they soon disappeared back underground.) Our three ant species were not a lot for Utah, Don announced, where 165 species exist.

"There might be some more small stuff in here, but it's hard to see with all this mud."

We walked back to the Malaise trap and Don unscrewed the collection bottle. At the picnic table, he poured the alcohol and insects into a large plastic container.

"My god!" I said. In the bowl was a wet, soft amalgam of wings and legs and plump abdomens, swimming in color—black and red and yellow and orange. "They look like jewels!"

Our heads bowed over the insect soup as Don used his tweezers to sift through and separate them, calling out their names.

"This one's an ant lion—it's a tiny mayfly-like insect."

The field guides I found at the library said: *Ant lions resemble damselflies but have longer clubbed antennae. One pair of wings extends horizontally and a second pair parallels the long body. Adults eat a wide variety of soft-bodied insects like aphids.*

"Oh, and you've got a spider—he must have crawled up there. And these are crane flies." They were pale dark yellow with filament-thin legs and antennae, much smaller than the one who flew indoors with me.

The adult crane fly resembles an oversized mosquito, but it does not bite people. The adult lives just 10-15 days.

"And we've got lots of these Hemiptera. I bet you know these."

He picked up a red-and-black bug. I leaned closer. "Box-elder bugs?" The traditional hints of red beneath the black wings had unfolded in the alcohol and bloomed to reveal a bright red thorax. When these bugs swarm around the yard or get into the house, I considered them mildly annoying; here, they were stunning and beautiful.

Box-elder bugs are true bugs that feed mostly on maple seeds. They may form large aggregations while sunning themselves; they don't get preyed upon in these large groups because they release a pungent and bad-tasting compound when disturbed.

"A couple of moths—here's a bigger one and there's a second one that's slight. And here's a honey bee, and this one is a large solitary bee—maybe a carpenter bee."

I recognized the orange and brown stripes on the fuzzy honeybee; the larger bee was twice as thick and wide. "Is that another bee?" I pointed to the far corner of the bowl.

"That's a wasp. And over here is a tiny parasitic wasp. You've also got a couple of syrphids." He held up a small black-and-yellow, striped, bee-like insect that's actually a fly.

Syrphid fly adults feed mainly on nectar and pollen. The larvae prey on aphids and other plant-sucking insects.

"Who are all the tiny guys all over the bottom? There must be hundreds of them."

"Those are midges—they're a kind of small fly, like gnats."

Adult midges often hatch synchronously with changes in weather patterns and form swarms and live seven days or less. They are an important food source for fish and swallows.

We trapped about eight kinds of flies, ranging from the tiny-tiny midge to house-fly-size and in-between. Don couldn't ID the species without a microscope, but there was a parasitic fly with a hairy abdomen, a tachinid fly, and several phorid flies.

Phoridae are a family of small, humpbacked flies resembling fruit flies. They can often be identified by their escape method of running rapidly rather than flying, which leads to an alternate name, the scuttle fly.

Our last insect was a diminutive lady beetle, tan and black, that I squinted to see. "I didn't know ladybugs came that small, or without red or orange," I said.

Don opened clear plastic bags and divided my insect catch among them. He added more alcohol, squeezed out the air, folded over the top stiff yellow band, then whirled the bag over the yellow band several times to form a seal. "Whirl-pack bags," he said and handed them to me.

I accepted the packets of shiny gems. I was amazed at who was living in my yard and in my bushes, and a bit remorseful that I did not know these neighbors—not just their names but even that they were living near me. The number of these tiny neighbors dwarfed the human ones on my block—both in number and in species. And these hundreds made a lot less noise.

"So this seems like a lot of insects, I mean for a typical yard, but is it?" I asked.

He shook his head. "Not really. You've got a lot of grass, which doesn't support much insect diversity. And if your neighbors use chemicals on their lawns, that kills lots of insects."

Even without the pesticides, an urban environment has fewer insects, which may seem like good news to some, but it's not that simple. Urbanization (with its prodigious pavement and buildings) fragments insect habitat and may increase the densities of undesirable insects while decreasing beneficial predator insects. Unfortunately, the conservation of insects is not a priority for most urban dwellers, even among those who generally appreciate and value the environment.

When I bought this house, the yard was 90 percent grass, essentially a monoculture. The insects who take it in the shorts in a yard like mine are the specialists and uncommon insects. An insect needs different plants and plant-cover at different stages of its life. The South African entomologist Michael Samways calls this polymorphism: the caterpillar is very different from the butterfly it becomes, and the habitats for both must be equally conserved. Creating a polyculture of plants increases natural enemy activity (code for predatory insects, like the ant lions and syrphid flies whose larvae eat aphids) and reduces any need for insecticides. The good news for me is that promoting insect diversity in my yard is pretty simple: reduce the grass, plant a wide range of plants (including natives), and don't poison anyone.

For some, a polyculture, wilder yard is a tough sell. Two University of Michigan scholars, Carol and Mark Hunter, concluded that highly manicured gardens (and parks) have persisted for over four thousand years in part because "scruffy nature" is unappealing to many people. To some, a profusion of native plants looks "messy" next to trimmed lawns and tidy flowerbeds. They recommend designing with "cues of care"—a narrow, mowed turf strip between wild, exuberant patches. The bit of grass is evidence that someone tends it, and the plant diversity is a boon for insect conservation.

A strategy they recommend for insect biodiversity in urban areas is installing natural (and low-maintenance) plantings along roadway easements, in storm-water management areas, and even on green rooftops. But they admit that urban insect conservation will never be on par with conservation of more charismatic species such as birds. It was my love of birds that led me to think more about insects, so perhaps that is a connection that can be strengthened.

~§§§§~

As Don packed equipment in his car, he reflected, "We have no idea how most insects make their living or their life histories, even if we can name them." Nor, about the relationships among insects. When Don studied one specific ant species, he discovered that when one of those ants dies, one kind of fly lays its egg only in the head of the dead ant. Another fly species lays its egg in the abdomen of the dead ant. And a third fly species lays its egg in the leg-elbow of that one dead ant. "Now that's an interrelationship," he said.

I am clueless about the complex interactions and niches for the invertebrates we captured. And, I have been guilty of stereotyping them: one grasshopper seems like another, a fly is a fly. When I bought ladybugs at the garden store to reduce the aphids in my yard, I thought, ladybugs are ladybugs. Pyle told me that what I bought were probably Asian ladybugs, which unfortunately prey on native ladybugs and butterflies. Of course: why would ladybugs in Asia eat the same things as ladybugs here?

Samways put a number to those interactions: "A thousand [insect] species, for example, in the same community (not an unreasonable figure) potentially produces 0.5 million interactions." It reinforced what I learned at the beginning of my insect journey: the amount of ignorance about insects is phenomenal. As Samways concluded, "Without insects, the likelihood is that the world as we know it would be radically changed in a matter of days." Insects aren't just important for biodiversity and the world's functioning; they *are* biodiversity itself.

In my reading, I saw quite a bit of research about how climate change is affecting invertebrate populations; like larger wildlife, some insects will move northward as temperatures warm or into new regions as precipitation and plant cover changes. (The current poster-insect for this is the mosquito that carries the Zika virus.) It is impossible to comprehend how climate-driven movements of even a hundred insect species will alter those 0.5 million interactions. As one scholar concluded, regardless of how climate change alters the animal kingdom, insects will remain the dominant life form.

Aristotle said we should study all animals without aversion, "knowing that in all of them there is something natural and beautiful." He proclaimed eons ago that everything in nature has its purpose, and there was an essential unity between the form (eidos) of an organism and its purpose or function (telos).

Insects' foreign form contributes to our ignoring or minimizing their function—these creatures of forgettable size of the spineless kingdom who skitter and fly with extra appendages, oversized compound eyes, and crunchy coverings over bloodless bodies. Yet, the amount of knowledge and function possessed by invertebrates on this planet is inestimable. They are the workers, the recyclers of decay, the conflict-solving colonies, and hierarchical families. They are the kitchen that feeds the food chain, the micro- and macro-plant managers of the world. They are small and foreign but larger than all our lives. We reside in the same kingdom Animalia. All I can do is give them healthy-messy places to live, and stand aside.

I sat in the backyard after Don left, the sun filtering through the fluttering maple leaves. How did we get here, so ignorant of something so remarkably and singularly important?

In most nonacademic writing I had read over several months, invertebrates were sharply divided into good and bad, like Oz's good witch and bad witch: they were one or the other. Bad ones were "pests" and called for chemical demise. They were good only if we were aware that we got something from them: honey, beeswax, silk, pollination of fruits and vegetables and crops, aeration of the soil, decomposition, insects for fish, beauty from the butterflies. Such utilitarian thinking is common in our species; even *Audubon* magazine titled an article, "What Do Birds Do for Us?"

But perhaps a larger reason we got here, why we so marginalize and demonize the insects among us and brush them aside, is our basic insecurity as a species. Samways said insects make us feel like *we* are the small and helpless species; because there are so many of them, they are "lilliputians in our much larger world." In his essay "Going Bugs," James Hillman wrote, "Imagining insects numerically threatens the individualized fantasy of a unique and unitary human being. Their very numbers indicate the insignificance of us as individuals." How ironic that the smallest, everyday creatures force us to face our biggest fear: humans do not run the world; it's the little guys.

After I spoke with him, Pyle graciously sent me a chapbook of his poems. I read "Letting the Flies Out" several times.

Every day I visit the half-screened
back porch to let the horseflies
out. Smart enough to get in,
too dumb to get out, they wham!
against the rusty screen
all day long.

The horseflies are big and slow and easy
to catch and toss out. The hover flies,
passive and happy to go. The smaller ones
have to fend for themselves. The bumblebees,
I need a glass for them.

Why release horseflies? They're no skin
off my nose, or arm, or neck,
if I watch out. Besides, I can't stand
frustration in any animal,

and a big fly battering a screen
is the definition of frustration.

And, oh! their stripe'd silken eyes
are beautiful.

⊙ ⊙ ⊙

⊙ ⊙ ⊙

⊙ *10* ⊙

Robotic Iguanas

On a ledge in a cliff face, a small brown iguana raises his head and says to the chartreuse iguana perched above him, "So, you wanna piece of me or what?" Soon the iguanas, joined by some toucans and macaws, burst into song: "Right here in the jungle mon, that's the life for me." Lights strobe, the toucans flap their wings, and the iguanas bob their robotic heads. On cue, waiters and waitresses join in the singing, waving strips of brightly colored fabric. A toucan shouts, "Hasta la vista baby!" as our waiter leans over and asks, "Who had the shrimp?"

Coming here was my idea. I had returned to Salt Lake City from a summer of outdoor experiences while millions of my fellow humans tuned into *Survivor* and hundreds stood in line at the new Mayan theme restaurant in the suburbs. A combination of stifling late August temperatures, unhealthy levels of smoggy ozone, and thick smoke from dozens of wildfires raging in the West left me feeling lethargic and listless. I called a colleague. "Chris," I said, "as people who teach environmental communication, we really ought to check out this restaurant with the Mayan jungle theme." To my surprise, she did not hesitate.

The Mayan is not the first nature-themed restaurant and will not be the last. The Rainforest Café and others have recognized that greenery and nature icons brought indoors create atmosphere and contribute to a brand. A trend-hunter website called it "bio-philic dining." Nature restaurants are another way that culture uses "nature" to sell products and colors our perceptions of what and where nature is. It's true that even wee bits of time spent in living nature have healing effects, but restaurateurs mainly desire getting crowds through the doors. In that regard, the Mayan is over the top with a show of robotic jungle animals, cliff divers, and performing waitstaff.

Our table is on the third level of the restaurant next to a railing with a good view down to the "stage," a cliff face about two stories high adorned with tropical plants made of plastic. A gentle waterfall pours from the cliff into a large aqua pool. The rocks are molded concrete and the pool reeks of chlorine. Under clear resin, our tabletop bears a colorful design of the Mayan alphabet and calendar. Down one level to the right is an area with carpeted steps and a large video screen playing cartoons—a sideshow for children not sufficiently captivated by robotic iguanas. High above in the middle of the cliff wall is an office window, light seeping from behind closed blinds.

The house lights—already dim—grow dimmer. On the lower cliff face, steam starts pouring from holes in the rock and two red eyes begin to glow, eventually illuminating a large fiery face in the stone. A deep voice booms, "I am Copac, behold the power. . ." The message is foreboding and a bit evil, something about heat from the center of the Earth. To break the tension, one of the toucans announces that it is about to get a lot hotter. The waiters and waitresses agree, chiming in with a chorus of "Feeling hot, hot, hot!"

Chris and I laugh; we are seated under an air-conditioning vent. By way of contrast, I tell Chris about a Guatemalan jungle I visited, and how even at 3 a.m. lying perfectly still, sweat would trickle from my face into my ears and hair. As the "hot, hot, hot!" number winds down a macaw asks for a cold towel. "I feel Mayan and I'm not even tryin'," it squawks, instructing diners to order another drink.

The owner of the Mayan—a quirky Mormon guy with an Intermountain empire of car dealerships, an NBA team, mega movie-theater complexes, and the new theme restaurant—said in a newspaper interview that he took great pains to give his patrons an authentic experience. He sent his architects to Mexico and Central America to ensure that the restaurant could recreate the experience of visiting an ancient Mayan community. They returned with proposals for plastic banyan trees, thatched huts covering computerized cash registers, and chlorinated water-falls. According to a lawsuit, what Salt Lake's Mayan restaurant allegedly re-created was not an ancient Mayan community but a nearly identical Mayan theme restaurant in Denver.

The lights dim again. A disembodied female voice speaks soothingly about standing on sacred ground, hidden in the jungle. Two young men in loincloths and tall, feathered head-dresses emerge on an upper cliff ledge and bang on tall drums, the slap of their hands occasionally out of step with the drum-beat of the amplified soundtrack. An image of a young woman appears on the rock wall, a water goddess of sorts with bright red lipstick. Her name is Tecal. "The spirit of the jaguar calls and I awaken," she says, urging us to return to a lost paradise, to the Earth, to celebrate, rejuvenate, and rebirth. As her speech cre-scendos, lightning flashes, thunder booms, and the once-placid waterfall gushes noisily into the pool, spraying the plastic ferns but not the diners beyond. People stop their conversations and turn toward the water.

Chris and I compare notes on our food (her taco salad with iceberg lettuce is unexceptional and my shrimp are tough) and discuss the "flood." She recalls how, in 1983, abundant moun-tain snowfall and an abrupt spring melt sent City Creek roar-ing through downtown. Like many such floods, it was caused by a combination of weather events and failed human attempts to control and divert runoff. Both that flood and the Mayan one, I point out, demonstrate a similar human desire for (and belief in) control of natural elements that by their very nature are largely uncontrollable and highly unpredictable.

The warriors return, wearing only Speedos and asymmetrical face paint. The crowd has been anticipating this, the most talked-about part of the show. From the highest point on the cliff, the young men alternate fancy dives into the pool, swim around the side of the cliff, and disappear to reappear at the top for another dive.

Shortly after the restaurant opened for business, the *Salt Lake Tribune* did a profile of one diver, who like all the divers was a member of a high school swim team in the valley. The diver they interviewed had emigrated from Guatemala when he was eight, and it was suggested that perhaps he had some Mayan blood in him. The story also mentioned that some restaurant patrons have asked whether the cliff divers are fake like the cliff they jump from and the lagoon they land in.

When I was a naturalist in Olympic National Park, I was frequently asked whether the deer wandering through the parking lot snarfing up Cheetos were real. Tourists also asked me what time it would rain in the coastal rainforest, as if there were a button we pressed, or as though, like Old Faithful, we could predict the rain. (Their interest was not so much in the natural patterns of a rainforest but in not getting wet.) Although the gulf between the real and artificial is vast, we have accomplished the illusion of no gulf at all, like silk flowers you must touch to determine whether chlorophyll lies within. We remake and remodel the natural world and its elements into more predictable and controllable versions, our own little theme-park paradise where flowers never fade on the vine and bubbling brooks never run dry.

There is more than humor or sadness in this degree of disconnect; there is danger. We grow increasingly ignorant of the natural original and risk not valuing it—or valuing its replacement more. When Salt Lake's foothills are abloom in early spring with allium, balsamroot, vetch, and sego lilies, most residents are aware only of imported tulips and daffodils. Numerous western cities have gone so far as to codify the imports, making green grass and thirsty flowers not just a cultural imperative but a legal one.

"You wannanother margarita?" asks our waiter.

"Is that what the Mayans drank?" I reply.

The skinny young man with spiky platinum hair stares blankly.

"No thank you," I say.

He leaves a small, black notebook on the table and says he'll take it when we are ready.

While paying up, I wonder if my fellow diners believe they are experiencing nature or just some wholesome family entertainment that comes with a mediocre meal. Can someone who knows only cultural and stylized depictions of the natural world—Disneyland,

PBS, the Nature Company—ever love and understand the wonderfully complex original? Can we care deeply about the jungle or the foothills or an untamed mountain creek if we have never truly known them? The love and compassion I have for the West is rooted in decades of discovery: reading the summer sky for signs of late afternoon storms, welcoming birds who pass by each spring, brushing against a thousand sagebrush to know its potent perfume. Such experiences remind me that my control is minuscule, my volition matters little, and the capacity for wonder and entertainment is infinite.

Chris and I contemplate what the Mayan has to teach us. Jungles are colorful, comfortable, and sublime. Ancients gods and goddesses—some benevolent, some not—control the weather. Nature is high-tech predictable and engineered friendly. Animals (with human voices) are there to amuse us. And you can buy pieces of the jungle in the gift shop to take home.

At the Mayan, the jungle and its creatures are most entertaining (and I guess, profitable) when they act like us, and for us. In the end, the Mayan is not so much about nature as it is about a culture that prefers plastic picket fences over wood ones, robotic animals over wild ones, reality TV over real life. The Mayan's jungle is commodified: a consumable experience where nature's qualities are appropriated and ritualized in indigenous people, stuffed parrots, and waterfalls on command. Such a material way of "knowing" the jungle neglects the history of colonialism and corporatism and the diminishment of nature there. This jungle is more fun.

The lights on stage grow bright and the birds start jabbering again. One introduces herself as Margarita Macaw; another, Pierre, wears a beret and says he is from Paris. Iguanas Marvin and Harry ask the macaws if they've seen their sunglasses. A bird informs us that "it's always perfect weather" in the jungle.

We leave a tip on the table; the entire show is beginning again.

⊙ ⊙ ⊙

⊙ ⊙ ⊙

⊙ *11* ⊙

Shut the Door

We sat on opposite sides of a long Formica table. Mr. Stern-feld, Chief Eating Officer at Mill Creek Fresh, was not happy with me.*

"If the door's not open, customers won't know we're open," he said. He explained that the open doors that seemed to cause me such outrage were not really open to outside air anyway, because they had a curtain of air from a fan that separated interior air from outside air.

"I understand how these fans work, and they're not all that effective and use quite a bit of energy to operate," I said. Silence.

In most all regards, I am not a confrontational person. But I kvetched often enough why I boycotted this store that I finally wrote the store owner. My bus to campus traveled in front of his store, an old gas service station converted in the 1990s to a specialty market. The parking sucked and prices were steep, but the neighborhood loved it. With just a handful of exceptions each year, one or both of the large glass-pane service bay doors was opened each morning, leaving a clear view inside of the heirloom produce, cheese, flowers, and artisan vinegars and jams. Every time my bus passed the market— 15 degrees one snowy morning or 99 on a sweltering August after-noon—those open doors rubbed my energy sensibilities raw.

Mr. Sternfeld told me that Mill Creek Fresh was an E2 Business, in part because they had dramatically reduced their energy use.

I bit my tongue; the city certification program granted the E2 title to businesses that simply changed cleaning products and a few light bulbs.

*Names have been changed.

"Our customers depend on our open doors to know when we're open," he said again.

"You've got a big and loyal customer base that probably knows your hours, but I'd be happy to do some survey research for you—free of charge—to better understand their sentiments about the open doors."

He declined.

The letter I had written to Mr. Sternfeld was constructive (and I thought persuasive), but in hindsight was more maudlin than necessary. (Okay, I cannot believe I invoked polar bears, but I kept visualizing the energy drifting out the wide doors, winging north, melting the sea ice, and drowning the massive white bears.) He replied quickly to my letter, shocked that I had so misread all the good environmental things the store was doing. He wanted to talk more about the "great food we sell" and discuss ways "we may reduce our energy use." But at the meeting's conclusion, it was clear he had no intention of changing the "feel of an open-air market" and shutting the damn doors.

"Why does this bug you so much?" a friend asked after the meeting. "Yes, he's wasting energy, and even more with all the exotic food he ships in, but a lot of businesses waste energy."

It was true; I knew plenty of other open-door shops and energy wasters. But somehow, I expected this local businessman who used all the right sustainability and green-foodie lingo to be different. The clever words in his radio jingles on public radio did not match what his wide open door communicated: profligate energy waste. And it did not match the gestalt of my childhood. When (Depression-era) Dad heard one of us kids fly through the family room door, he would yell from the next room, "Shut the door! We don't live in a barn!" He also threatened to dock our allowance for leaving on lights (though he never did). Waste was waste; energy was valuable, then and now. And even more so now.

And truthfully, that open door bugged me as a communication professor who thought she could communicate her way to a good solution. Since my written appeal and our interpersonal conversation weren't successful, I needed to investigate what the communication research said might work: perhaps providing scientific information about energy, appealing to local pride, or maybe invoking customer attitudes? Heck, I was ready to stand outside his store with protest signs and call the media. Somewhere was the magic message and motivator that would shut the damn doors.

Thus commenced the Sternfeld Project, my quest through dark halls of psychology, economics, sociology, and engineering to move Mr. Sternfeld to action. But over the years (yes, years) of the project, what I really learned was how all types of energy are tightly bound to cultural values that condition who we think we (and others) are. The culture of energy was far more powerful than I realized.

꿁꿁

I began the Sternfeld Project where I was trained to begin: what did past research find most successful in getting individuals to conserve energy? Most of the studies focused on individuals' energy conservation behavior at home, though this direct use (such as turning on a laptop) pales in comparison to the energy we use *indirectly* in goods we purchase (laptop manufacturing uses forty times the energy than the device consumes in a year of use).

The studies (conducted mainly in North America and Europe) tested variables both demographic and psychological, and employed theories and models galore. Many studies used the well-loved Theory of Planned Behavior, which grew out of the Theory of Reasoned Action (and when I put my nonacademic hat on, I find those names hilarious). In an undergraduate research class, my students and I tested this theory in a survey of drivers along the Wasatch Front. We designed questions to test what might influence the target behavior (less driving), such as their past behavior (did they ever walk instead of drive) and their attitude toward the behavior (did they believe reduced driving was important). We asked about their value orientations (environmental or otherwise) and about their knowledge of alternatives to driving. We asked how much control they thought they had over the amount they drove. And finally, we asked about their intention to perform the target behaviors: would they take steps to reduce their contribution to air pollution or participate in a campaign that promoted walking and smart driving.

The students got excited because people said yes; many were indeed interested in taking these steps. I did not tell the students that getting people out of their cars—for whatever reason—is the hardest behavior change conceivable. Saying *yes* to a student surveyor is very easy. And every year, more drivers drive ever more miles along the Wasatch Front and use more energy.

My friend Carol resolved one day to ride the bus in her Florida town during the coming year. And every year I ask her if she's ridden it yet; no, but it's still on the resolution list. I tell her we need "bus ed" in schools like we have "driver's ed."

There were no easy answers in the home energy conservation studies. The studies identified a laundry list of variables associated with energy conservation: income, age, education, home ownership, desire for comfort, and incentives. Increasing income and increasing house size were associated with increasing energy use (which my students might call a *dub*). And, people underestimated the amount of energy they used by a factor of three.

Even if people felt positively about using less energy, it often did not happen. One research team noted multiple constraints, including financial. One campaign to reduce energy use across an entire community didn't achieve significant reductions; evidently peer pressure wasn't strong enough to change what you did behind your front door. Another study concluded that a sense of powerlessness prevented people from turning their positive environmental attitudes into low energy use, and a desire to self-indulge sometimes won out. The bottom line: changing individual energy use is a crapshoot, even though it's the approach that researchers and campaigners take most often.

What if you made the use of energy (a largely invisible commodity) more visible—would that help people use less? Several research teams did just that: they installed monitors in people's homes that displayed energy use in real time. Yet in most cases, the new monitors barely budged consumption. In one experiment, the Belgian researchers concluded, "We have observed that the metre can change electricity perception, but that only households already interested or involved in energy savings are willing to use and learn with the monitor. We suggest that these devices should accompany a deeper transformation of the 'culture of energy.'" They concluded, "We have to invent other ways of *making energy precious* than making it visible through a small monitor" (emphasis mine).

When I stand at my bus stop in the morning, I watch a guy across the busy four-lane street come out his front door, start his car, then smoke a cigarette while standing on his stoop. The house is rather shabby; the front picture window often fogs and freezes with moisture. I often debate going over and talking to him and just catching the next bus. I've written the script in my head dozens of times. Did he notice our air pollution, and did he know that the first five minutes of driving (or idling) are the most polluting? Did he listen to the Car Talk guys say that modern cars need no warming up at all—just GO? Did he know that Salt Lake City has a law against idling and he could get a ticket?

My script is predicated on *information* being able to change his mind and make him see the light—the classic "information deficit model" that researchers simultaneously reject yet continually return to. Since the guy's behavior is the same most every morning at this same time, it's his ritual, his habit. Maybe he's got a bleak life. Maybe getting into a warm car is one of the few pleasures he has. Maybe he's trying to be efficient by having a smoke while warming the car. He smokes, so maybe air pollution in his lungs does not bother him. Could I appeal to him as a neighbor who lives a few blocks away, as a neighbor who is harmed by this air?

My friend Camille and I talk about how we might intervene when we see idling cars. We feel strongly but are loathe to confront. She thought about handing drivers a piece of paper with her request and quickly walking away. We laugh: confrontation-light. I do confront my bus drivers; the bus stop on campus is the end of the route, and next to the bus in black letters is a "No Idling" sign. The engines putter and hum while the drivers stand on the sidewalk on their phones or having a smoke. "Did you know this is a

No Idling zone?" I ask and point to the sign. "Huh, didn't know that." Another replied, "Whatever." One driver said he had enough to deal with just with the crazy drivers. I have reported the idling to the transit company, but the behavior does not change.

When individual behavior harms the collective good, it's called a collective action problem. I receive no immediate benefit (in air quality or the upper atmosphere) from taking the bus, and the guy across from the bus stop experiences no direct harms from his actions either. It takes a mass of individuals taking positive action to solve the problem (which is often why and when government steps in). In California on a bad-air-day when people are "encouraged" to take mass transit, *more* people actually drive; they assume that others will bus-it and leave the roads less crowded.

After reading the reams of research, I was not feeling very hopeful about changing Mr. Sternfeld's attitudes and behavior. Shutting doors is not part of the "culture of energy," and energy is not considered precious. I needed to find other doors to open.

❦

One reason why Mr. Sternfeld's open door bugs me is how it models this wasteful behavior for others. We humans are strongly affected by what our fellow humanoids do, from fashion to language to energy use. We learn the "social norm" by observing what others are currently doing and/or judging what we think they ought to be doing. Social norms are enforced via social approval or social ostracizing.

The social norm now seems to be propping open doors and not thinking there is anything wrong with that. In Berkeley over Christmas, in Jackson Hole in summer—most all of the shops leaked hot air or cold air respectively onto the sidewalk. They must believe as Mr. Sternfeld: an open door shows you are open for business somehow more effectively than a neon OPEN sign.

At the Oasis Café in Salt Lake in late October, the double doors were propped open. I said to the host inside, "I gotta say I'm not happy about your open doors. It's not very warm outside and leaving them open wastes a lot of energy." He replied, "Yea, well, glad I'm not paying the bill!" and showed us to our table. Of course hours later I thought of a retort: "Actually, you are paying the bill."

I walked my dog past the Carter Baby store on a chilly spring day. "Excuse me," I spoke through the open doors, "but your open doors are wasting a lot of energy and I'd really appreciate it if you closed them." The slender woman said primly, "Well, that's just the way we do things around here" and returned to folding clothes.

Invoking social peer pressure can backfire, too. Several utility companies tried using the monthly bill to show how much energy you used in comparison with your neighbors in hopes of spurring some competition. It spurred competition, but in the wrong direction; when low energy users saw that some neighbors used far more, they consumed more, too.

A campaign in the UK called Close the Door has combated the issue on several fronts. They commissioned a study of businesses with closed or open doors and found *no* significant difference in foot traffic. The study also found that closing the door reduced energy use 20 to 50 percent. Another study found the "air curtains" were only 50 percent efficient (and less so every time a person walks through and disrupts the flow) and consume 24 kilowatt-hours per day to operate (the equivalent of running four very large refrigerators). Campaigners from Cambridge to York went door to door, handing businesses the reports and making their case. They gave them Close the Door stickers to advertise their participation, and emphasized energy leadership, saving money, and pleasing an increasing number of customers who cared about the issue. The Close the Door website lists some closed door success stories. Their multipronged approach included science, person-to-person contact, and appeals to values and economics.

New York City went a different direction to close shop doors: they legislated it. A law took effect in July 2016 that fines stores and restaurants $250 if they keep doors or windows open while running the AC.

Mr. Sternfeld, the café, and the baby store didn't respond to my pleas; would they respond to a law? Perhaps, but a law is only effective if it's enforced, which the local idling law is not.

At this point in the Sternfeld Project, I was feeling discouraged and a bit like Donna Quixote charging around on a horse with a lance. How could I focus on open doors when there are outdoor patio heaters heating the doorless great outdoors for our "comfort"? When half of the hallway lights in my office building can *never* be turned off (they told me the building code mandated it for "safety"). When the new default temperature on a new thermostat is 72 degrees (what happened to 68?). When huge holiday lawn decorations require constant electricity to stay inflated (including the ironic polar bear one block away). Even public bathrooms are an entirely electric experience: water, soap, towels, toilets. My house at night has a green glow from all the beaming gadgets that never sleep. If energy is doing and powering all these things, why is the only time we think it valuable is when the power goes off?

And, did it make sense to target Mr. Sternfeld (or any individual) when things like thermostat manufacturers and bathroom designers are in the mix? Individual energy consumption (homes, cars, and so on) has never been more than a quarter of the consumption used by other sectors: commercial, industrial, corporate, agribusiness, government, military, and so on. And of course, it's all connected: the thermostat and glowing gadgets are used in my home, but I did not manufacture them.

A campus colleague and psychology professor, Carol Werner, recognized how embedded individuals are in a larger system when she presented the Holistic Model of Behavior Change to my class in a guest lecture. In the center of the model's several concentric rings lay the Individual, who was surrounded by the all-important ring Social Context of friends, family, social media, and so on. If energy use is not a consideration in an individual's social groups, it probably won't be for the individual. The next ring was the Physical Environment, which could include where your nearest bus stop was, the walkability of your neighborhood, types of outdoor lighting, and dedicated bike lanes. The outermost ring was Socioeconomic and Political, where idling and open-door laws get passed, where gasoline and transit are priced, utilities are regulated, and thermostats are sold. It's easy to see that all these levels need to work together for energy conservation to be effective. Individuals are the small fry, swimming in the soup of our energy culture. It makes sense to consider other and bigger fish.

But what happens if there isn't congruency or agreement among these levels that energy is *precious* and that its use in our everyday lives has immense local and planetary consequences? Where do you start then? I shifted my reading to the culture of energy and the nature of energy itself.

Energy is the capacity to do work. As I type this, my fingers express kinetic energy, my brain is engaged, my heart circulates blood, my lungs process oxygen. I got the energy to do all this from food, which got its energy from the sun. The sun is the original source of *all* energy, which the Earth receives in the form of light and heat. Plants transform this energy through photosynthesis into their growth and reproduction, and animals who eat the plants transform the plant energy in their guts into kinetic energy.

Of all the ways I used energy this morning, my brain first thinks about electricity: the stove that boiled water for coffee, the microwave that cooked oatmeal, and the laptop plugged in. But in truth, everything around me contains that original sun energy transformed into matter: the metal of my coffee mug, the clothes on my body, even the dog herself. Transformation of energy from one form to another is what life is all about.

Most electricity is generated from fossil fuels, which are ancient fossilized sunshine; plant material was compressed over eons to form coal, oil, and natural gas. Even hydroelectric power depends on the sun's light and heat to power and churn the planet's hydrological cycle and fill those rivers with water.

I worked with a master's student whose thesis project traced the electricity in his house back to its various sources. Ross began his series of radio stories by asking people on the street where they thought their electricity came from. One person he encountered was a journeyman electrician who had no idea where that electricity came from.

The invisibility of energy is due in part to the infrastructure that delivers it. Electric lines and buried gas lines sanitize energy and divorce it from its sources, much like plumbing divorces us from sources of water. It's easy to take them both for granted—assuming lights will come on when we flip a switch and water will fill a glass when we turn the faucet. Another challenge is that energy is both hard to picture and hard to put in terms we get. We understand calories as a measure of food energy, or the number of trees needed for a ream of paper, but a kilowatt, what is that? However energy is measured, we know we can expend less of our own personal energy when we rely on external sources of energy.

꧁꧂

Food is one product that I and many others do not expend personal energy growing; we go to a store for most of our food. And in the last several decades, life has been good for us "foodies." My mother—a foodie before her time—would have marveled at the riches of foodstuffs on grocery shelves today: heirloom this and that, cheeses from around the world, a half-dozen colors of potatoes, seventeen kinds of balsamic vinegar, and olive bars. She would have been flabbergasted to buy fresh strawberries year-round, let alone fresh mangos, artichokes, and pineapple. She would have encountered the maze of egg-buying choices: cage-free, free-range, antibiotic-free, hormone-free, omega-3, vegetarian, soy-free, natural, pasture-raised.

This cornucopia was delivered to you by cheap fossil fuel energy from countries all over the globe—and all year long. Those summer oranges look wonderful (Australia), and those winter grapes are so fresh (Chile). Mr. Sternfeld has been involved in Salt Lake's "local food movement," which is a good thing, and he proudly labels his local lamb and local cheese. But as a foodie market, his store (and all stores) participate in the cultural norm (and high energy costs) of season-less food (fresh flowers year-round and tomatoes far beyond the local growing season) and a global food culture (exotic produce and condiments shipped from Brazil, Denmark, Africa). It's "how we eat" now—that is, it is how wealthy people eat in a developed country when they have cheap fossil fuel energy.

Modern agriculture is an extremely energy-intensive business. The U.S. Energy Information Administration reported that the U.S. agriculture industry used nearly 800 trillion British thermal units (Btu) of energy in 2012, or about as much primary energy used in the entire state of Utah. But only 20 percent of the energy used in the food system is needed to grow it; the remaining 80 percent of energy is used to process, package, transport, refrigerate (retail and home), and prepare food. Because the food industry is increasingly globalized and concentrated into fewer, large suppliers, that means food is traveling farther to get to us—what some call "food miles." The average distance fresh produce travels is about 1,500 miles (1,200 miles for processed food). That's staggering, but food miles only represent about 10 percent of the energy costs of food. Equally staggering is that Americans waste about 20 percent of the food they buy—similar to the percentage of energy wasted from open doors.

~~~~~~

Because every piece of food (and every auto and every cell phone) was transformed from sun energy into something else, it's easy to see why the energy industry is such a central and powerful piece of our economy and our culture. It's a Keystone industry. A society's sources (and choices) of energy make it possible for all other industries (like the food industry) to operate, which makes it possible to produce and practice our culture. Throughout human history, we have searched for new and additional sources of energy not just to survive but to create surplus production and population and to sustain our broader social, political, and aesthetic culture.

In a fascinating book, *Art and Energy: How Culture Changes*, Barry Lord argues that each new type of energy (wind, water, slave, or coal) brings with it certain cultural values that condition who we think we and others are. Each energy source requires us to adopt some values (and suppress others), set priorities, and make sacrifices. Take, for example, "slave energy"—yes, many societies used slaves as energy, which greatly extended the "capacity to do work." Living in a slave-owning society meant you had to accept (whether you approved or not) the premise that some people were entitled to own slaves (and that slaves had no right to control their lives or the lives of their children). If you depended on this source of energy—if you were constantly calling upon slaves to do things for you—it conditioned the kind of person you were. If you did not own slaves yourself, you still benefited in the market-place from slave energy, which conditioned you. Similarly, coal as a primary energy source conditioned the work lives and work ethic of coal miners; it was hard, dangerous work that required a highly disciplined, communally organized, and class-conscious mass workforce. By comparison, oil and gas extraction requires a relatively small number of workers, which shifts the locus of value from production at the wellhead to the market (and marketplace).

In other words, Lord says, energy sources are not value-neutral. Since we require an energy source to practice our culture, we accept the terms that accompany that energy, often without acknowledgment or debate. We are so conditioned by the availability of gasoline to "run to the store" (a phrase that confuses foot power or horsepower with fossil fuel power) that we have strong expectations and a sense of entitlement that this energy source will serve our immediate desires and what matters to us (convenience, independence, freedom to travel at will), values that have become integral to modern society. In very many ways, we accept (or justify) the terms of its use: air pollution and the attendant health costs, massive amounts of concrete, traffic, noise, climate change. Lord says our priority for each new source of energy seems to be efficiency and cost; impact on the environment is a secondary, if not tertiary, concern.

Coal (when it was the primary energy source) fueled a culture of production (think of the giant steel mills) where a culture of workers strongly identified with their jobs: "I am what I make." By contrast, Lord argues that cheap oil and gas fostered a culture where consumption was valued for its own sake, and workers identified less with (and were less tied to) their jobs and identified more with their possessions: "I am what I buy." Despite political, religious, or economic differences, other countries look to America as the leader in the culture of consumption.

Lord traces how the automobile drove the oil era, and how credit was the way to transform everyone into a consumer even if he lacked money. Henry Ford let customers buy cars with only a small down payment, and oil companies began issuing "plastic" (credit cards) to fill up those cars at their gas stations. Department stores soon imitated with their own credit cards. Now, the average U.S. household's unpaid credit card debt stands at about $10,000.

A bubbly student in my large-lecture class liked to wear a pink sparkly T-shirt that proclaimed she was "Born to Shop!" It's a fitting symbol for the dominant value of oil and gas culture: consumption. It dominates daily life and changed the way we think about most everything, including ourselves. The cultural identity once provided by one's work and community has been supplanted by brand identity and seeing oneself and others first and foremost as consumers: I shop, therefore I am (even if I am in debt).

It's easy to see why the culture of consumption has parked a bulging shopping cart in the way of energy conservation. Only China consumes more energy than the United States, and much of their energy is used to produce U.S. consumer goods. Because of the core role of consumption in the growth mandate of capitalism, the government gets nervous (if not paralyzed) about making changes or placing restrictions on energy use, no matter the effects of consumption on the environment and climate change. It's worth noting that capitalism itself is a social construction created by our culture that requires us to accept and suppress certain cultural values.

〜ᕦᖾᕤ〜

One day, I got off the bus long before my stop to visit Mill Creek Fresh. A wood palette of asparagus and spring onions stood under the awning in front of the open doors. Just inside were many more palettes of produce, some labeled organic and local. Along the back wall rose shelves of olive oils, vinegars, jams, and condiments. To the right, the newer addition to the original gas station was bursting with case upon case of cheeses and meats. To the left were coolers of flower bunches and shelves with freshly baked bread.

"May I help you find something?" a young woman in an apron asked.

"No thanks, just browsing."

It's a sensorial shop that feels like a market more than a grocery. Your eyes feast even if your basket does not fill.

When I got home, I navigated the Mill Creek Fresh website. Under every tab, I read the words *authentic* and *integrity*, with a sprinkling of *nuanced, traditions, farmstead, intensity, purveyor, sublimely complex,* and *charcuterie* (I had to look up that one: cold, cooked meats). Though there is a focus on *local,* there is an unabashed emphasis on the exotic global. They boasted of traveling across the globe to find the best food the world offers. They wandered muddy pastures in one country and farm fields in another. They ate at home kitchens in various continents and knew all these providers personally. All the growers cared about their foodstuffs and knew the pleasure of eating.

From the descriptions as rich as the food, I understood what Mr. Sternfeld was endeavoring to create: shopping as an *experience,* the ultimate metaphor for consumerism today. His "fresh market" experience starts at the open front doors, evoking Seattle's Pike Place Market or a California produce market, as much as a European market. One customer said she hadn't had such good cheeses since she lived in Europe; another customer appreciated the goose fat which made her recipe as good as she remembered in France. The consumer experience he has created is worldly, elegant (some might say snobby), and attempts to make shopping feel like part of a yesteryear farm-to-table relationship. His food-topia vision does indeed matter to the planet because it is a très impactful use of energy (air curtains, massive food miles, meat), which sadly remains invisible to many, just like our energy use everywhere.

⁓≈≈≈

Even if I rail against energy waste wherever I see it and commute by bus, and even if I want to step away from fossil fuel energy tomorrow because I am intensely concerned about climate change, I cannot. I must use it if I want to do anything at all. It is hard for an individual living within our dominant energy system to question and disrupt the system itself; it is far larger than the energy; it is the entire system and the consumer culture that energy produces. What the Sternfeld Project taught me is the enormous scope of the challenge, the entwined existence of fossil fuel energy and cultural and consumer values. In a way, that is equal parts comforting and disturbing. I now understand why denial of climate change seems self-protective to some (who fear changes to the current system), while to others it seems the epitome of urgency and inaction is sheer insanity.

My students eagerly discuss switching to renewable energy, and a handful of friends have installed solar panels and drive hybrids. This is helpful and communicates much, but panels on every rooftop and hybrids in every garage will fall far short. The shift must include a profound recalibration in how we see ourselves and each other, as well as a dramatic change in our relationship to the planet that sustains us. This includes a shift away from consumer identity of "I am what I buy" to perhaps "I am what I save and protect." It's a deep, difficult cultural shift.

In the 1500s, Niccolò Machiavelli wrote of the arduous task of the reformer:

> There is nothing more difficult to carry out, nor more doubtful of success, nor more dangerous to handle, than to initiate a new order of things. For the reformer has enemies in all those who profit by the old order, and only lukewarm defenders in all those who would profit by the new order. This lukewarmness [arises] partly from . . . the incredulity of mankind, who do not truly believe in anything new until they have had actual experience of it.

When abolitionists called for an end to slavery, they encountered enormous economic and political resistance. People argued that the institution (and the energy slaves provided) was too large to dismantle and the economic impacts of doing so too great. The power of the vested interests (who passed along the system's costs to third parties, which included slaves) was tremendous. Although there are important and substantial differences in the comparison, transitioning from fossil fuel energy threatens entrenched and powerful interests who resist change—change that will perforce change everyday life as we know it.

A society's shift in energy sources is indeed, as Lord con-cluded, an engine of cultural change. Change is possible when new generations adopt new cultural values that an emerging source of energy makes possible. It gives me hope to remember that cultural change is constant, and that there are a multitude of positive, hopeful values ready to be adopted: ecocentrism, com-munity, moral responsibility, and less carbon, less consumption, less waste. The need to transition from fossil fuel energy may very well be the crisis that will change our relationship with the Earth—from one of sole entitlement of all earth's "resources" and of human gratification above all else, to a sense of earth citizen-ship and a large community of beating hearts.

So, will I end the Sternfeld Project? I'm not sure. While I realize the issue is much larger than any one store's open doors, I also need to believe that this one simple relationship—between me and a businessman in my neighborhood—is very important, and that what this one business models in terms of energy behav-ior is of import, even though energy waste happen a million times a minute. I need to believe that individual reformers and their beliefs and actions make up a culture and help drive its change, which they continually recreate and reinvent. I need to believe—indeed I do—that communicating and making visible the uncon-scious and second-nature use of energy is crucial.

Someone said that it is not that people do not want to change, but they do not want to *be* changed. That is perhaps the best lesson of all for me. Invoking drowning polar bears and a sole focus on his wasteful doors was not a good approach. I could have asked him why the open doors were important to him. I could have sought to identify common values we shared about living on this planet and in this culture. We both could have listened with-out judgment. Perhaps then we could have arrived at a common understanding that all energy is inestimably precious.

☉ ☉ ☉

☉ ☉ ☉

☉ ☉ ☉

Epilogue

The Everyday Is Everything

Yellow yard signs dotting my neighborhood implore me to "Protect Wild Utah." I wholeheartedly agree about the importance of wild places and designated wilderness; they have always been a salve and tonic to my soul, and I cannot fathom my life without them. Yet I know that what threatens those wild places most begins right here, where I live. Breakneck consumption and population growth are shrinking and compromising all wild places, and even the best campaigns to protect the wild are not powerful enough to compete with that.

Everyday nature is where it all begins—our very lives, and our very real ecological crisis. We took and took earth elements from all that wilder nature to produce all the nature-made objects here in everyday life, which shrunk the wild we cherish. It is not enough to value nature far away; we must see and reimagine nature close by and its inestimable value.

As a person who has never fully felt like a "city person," this was my challenge, too. I needed to examine and truly see nature where I lived. I needed to unwrap cultural meanings that blurred how I saw everyday nature: animals, energy, and the quiet beauty of daylight and darkness moving through this cityscape. I learned that so many of my perceptions and valuations were shaped and delivered by others (social groups, media, and the larger culture), such as insects are enemies, and noise is something our bodies and minds just get used to. And happily, I learned that I could see and value nature here, like the countless, uncultivable interactions between my backyard silver maple, the birds, the insects, the air, and me. These are the foundational bits that I had not truly connected to my wilder haunts. I do see nature more clearly and in infinitely more places now.

I spent years taking this second look, this measured consideration of the everyday nature where I live. How might you, dear reader, more easily reimagine largely unseen and uncontemplated everyday nature and make it part of your considered circle? How might it gain a place in your language, in your thoughts, in your field of view, in your moral family?

First, see all nature without immaterial boundaries—the inside and outside, the living and the once living, the wild and the heavily peopled. Our ability to see and comprehend the spillover and the connections are our greatest hope—and I do not say this lightly—for the preservation of us and all our Kin.

Second, pay attention. "Every day I see or hear something that more or less kills me with delight," begins Mary Oliver's poem "Mindful." She discovers these moments not in exceptional or sublime pieces of nature, but instead in "the ordinary, the common, the very drab, the daily presentations" of it. It is in these everyday moments where delight and wisdom reside.

Modern culture has schooled us well in not paying attention for very long or very deeply: twitter and text, switch channels and websites, drive fast, multitask, Instagram. Being mindful is a gracious and graceful attitude where you truly see and sense what surrounds you, in this very moment. Peel and eat that mandarin to experience the everyday nature of nature.

Recently when preparing a holiday dinner, I found myself grinning with the sound of cranberries—pop, pop, puff—as they split their skins in the bubbling pot. Then, the crackle-crack of ice cubes dropped into a glass of water. And, the slanted afternoon sun illuminating the steam rising from the potatoes. This was all nature; it was all ordinary, and it was joyous.

When Eric Freyfogle, a professor at the University of Illinois College of Law, spoke at my university, he said we need to learn to talk and think about nature in different ways, to develop a *second language of the common good*. The strands, habits, and patterns of our culture simply do not fit well with nature and do not accommodate it well, he said. Due to the way we see ourselves in relation to nature, we tend to fragment it into pieces. We think in the short term. We think ourselves somehow more morally relevant than other species (*human exceptionalism*). We think that individual rights, liberty, and equality all relate to humans but not to nature. A second language of the common good, he said, would value interdependencies and recognize that we are co-members of a community who partakes of public goods (air, water, food, land, and each other), whether we acknowledge that or not.

Though one could accuse a communication professor like me of viewing the world through language glasses, a *second language of the common good* is a wonderfully descriptive term for what is essentially the language of ecocentrism, the language of *Words That Come Before All Else*, the language of ecology. And the *common good* extends far beyond *the good of humankind* to the healthy functioning of all systems on this planet for the good of all Kin. It puts all nature—which includes us—in one gigantic box.

Arriving at this belief of a macro common good seems an enormous shift for a culture that puts humans at the top of the heap and in charge of all below us. But believe me, the shift can start small, and when you make it, it's such a relief to let go—from Plant Manager to plant partner, from trying to control water to envisioning water as "wandering wildlife." And it enlarges to a colossal extent what you get to consider as your "family." As Alan Watts (a writer and interpreter of Eastern philosophy) wrote, "You didn't come into this world. You came out of it, like a wave from the ocean. You are not a stranger here." You live within a huge, vibrant family of Kin.

I have also learned in my study of everyday nature that I need (we all need) nature to be more fully present in our cities. For many of the world's people, everyday nature is all the nature they have, and, sadly, that nature is often extremely degraded and squeezed out. I recently spoke in Taiwan: near my hotel, a fog of air pollution, streets dense with buildings and without trees or greenery, cacophonous traffic, and a murky river contained by concrete. Cities the world over now face air pollution, that most common common-good of all—from Paris to Delhi, Fresno to Pittsburgh—which is part of the same fossil fuel dynamic that is changing the global climate and chipping away at the pristine wild.

Both at home and abroad, I see and sense a profound dis-ease, which I attribute in part to a disconnection of humans from the physical peace and emotional healing that healthy nature provides. It is a tragedy to which we all have contributed, even if we are lucky to live or have access to the gift of healthy nature.

Cities need to be quieter! Cities need to be darker! City nature should not be a token, an afterthought, a human-dominated and chemical-laden monoculture that bears no resemblance to the wilder spaces around the city (if they still exist). Space for city nature is as vital a priority as space for a hospital, a school, a library. City nature needs thorough consideration of local ecosystems, native plants and animals, and climate; that is not the same as turning an arid vacant lot into a green sea of grass.

If you are lucky to have a vacant space nearby, dream into it. How could it be a healthy, inviting space for humans and all our Kin? Get all your neighbors together. Find out what the space looked like a hundred years ago and what Kin lived there. Involve and engage; create ownership through sweat equity and discovery. Sure, put in a swing set, but also trees and shrubs and messy places to explore and hike. And sure, okay—cram it full of Pokémon creatures to find.

References

ACKNOWLEDGMENTS

Colman, David. "In Praise of Early Adapters," *New York Times*, March 9, 2008

PROLOGUE

Bennett, Jane. *Vibrant Matter: A Political Ecology of Things*. Durham, NC: Duke University Press, 2010.

Corbett, Julia B. *Communicating Nature: How We Create and Understand Environmental Messages*. Washington, DC: Island Press, 2008.

Cronon, William, and Michael Pollan. "Out of the Wild: A Conversation Between William Cronon and Michael Pollan." *Orion* (November/December 2013): 65-71.

Cronon, William. "The Trouble with Wilderness." In *Uncommon Ground: Rethinking the Human Place in Nature*, edited by William Cronon, 69-90. New York: W.W. Norton, 1995.

Franklin, Deborah. "How Hospital Gardens Help Patients Heal." *Scientific American*, March 1, 2012.

Latour, Bruno. "It's the Development, Stupid! or How to Modernize Modernization." *EspacesTemps*. Cited in *Vibrant Matter: A Political Ecology of Things*, edited by Jane Bennett, 115. Durham, NC: Duke University Press, 2010.

Ulrich, Roger S. "View Through a Window May Influence Recovery From Surgery." *Science* 224 (April 1984): 420-422.

CHAPTER 1

Abram, David. *Becoming Animal: An Earthly Cosmology*. New York: Vintage Books, 2011.

Bekoff, Marc. *The Emotional Lives of Animals*. Novato, CA: New World Library, 2007.

Corbett, Julia. "Talking with a Cougar." In *Listening to Cougar: True Tales, Tall Tales, Folklore and Field Stories of America's Greatest Cat*, edited by Marc Bekoff and Cara Blessley Lowe, 83-90. Boulder: University of Colorado Press, 2007.

———. "Communicating the Meaning of Animals." In *Communicating Nature: How We Create and Understand Environmental Messages*, 176–212. Washington, DC: Island Press, 2008.

References

Deming, Alison Hawthorne. *Zoologies: On Animals and the Human Spirit.* Minneapolis: Milkweed Editions, 2014.

Kellert, Stephen R. "Affective, Cognitive, and Evaluation Perceptions of Animals." In *Behavior and the Natural Environment,* edited by Irwin Altman and Joachim F. Wohlwill, 117-151. New York: Plenum Press, 1983.

McGuane, Thomas. *Some Horses.* New York: Vintage Books, 2000.

Safina, Carl. *Beyond Words: What Animals Think and Feel.* New York: Henry Holt, 2015.

Shepard, Paul. *The Others: How Animals Made Us Human.* Washington, DC: Island Press, 1997.

Wilson, E. O. *Biophilia.* Cambridge, MA: Harvard University Press, 1984.

World Wildlife Fund. *Living Planet Report 2016: Risk and Resilience in a New Era.* WWF International, Gland, Switzerland, 2016.

CHAPTER 2

Berry, Wendell. *The Unsettling of America: Culture and Agriculture.* San Francisco: Sierra Club Books, 1977.

"Bloom Recyclable Laptop." October 2010. www.youtube.com/watch?v=WQX_NGb5vXs.

Chamberlain, Jess. "The Zero Waste Home." *Sunset,* January 2011, 74-81.

Eatmon, Thomas. Allegheny College. sites.google.com/a/allegheny.edu/teatmon/.

Hawken, Paul. *The Ecology of Commerce: A Declaration of Sustainability.* New York: HarperCollins, 1993.

Geyer, Roland, Jenna R. Jambeck, and Kara Lavender Law. "Production, Use, and Fate of All Plastics Ever Made." *Science Advances* 3(7)(July 2017). DOI: 10.1126/sciadv.1700782.

Jensen, Derrick. "Forget Shorter Showers: Why Personal Change Does Not Equal Political Change." *Orion* (July/August 2009): 18-19.

Jordan, Chris. Photographic Arts. www.chrisjordan.com.

Kennedy, Greg. *An Ontology of Trash: The Disposable and Its Problematic Nature.* Albany, NY: SUNY Press, 2007.

Leonard, Annie. *The Story of Stuff.* New York: Free Press, 2011.

Moore, Kathleen Dean. "Concrete Footing: On the Solidarity of the Insubstantial." *Orion* (July/August 2012): 46-48.

Motavalli, Jim. "Waste Not." *E Magazine* (March/April 2011): 22-29.

Rogers, Heather. *Gone Tomorrow: The Hidden Life of Garbage.* New York: New Press, 2005.

Scanlan, John. *On Garbage.* London: Reaktion Books, 2005.

Strasser, Susan. *Waste and Want: A Social History of Trash.* New York: Metropolitan Books / Holt, 1999.

CHAPTER 3

Chamovitz, Daniel. *What a Plant Knows: A Field Guide to the Senses.* New York: Farrar, Straus and Giroux, 2012.

Cronon, William. "The Trouble with Wilderness." In *Uncommon Ground: Rethinking the Human Place in Nature,* edited by William Cronon, 69-90. New York: W.W. Norton, 1995.

References

Hess, Scott. "Imagining an Everyday Nature," *ISLE (Interdisciplinary Studies in Literature and Environment)* 17, no. 1 (January 2010): 85-112.

Kimmerer, Robin Wall. *Braiding Sweetgrass: Indigenous Wisdom, Scientific Knowledge, and the Teachings of Plants.* Minneapolis: Milkweed Editions, 2013.

Simms, Michael. *Apollo's Fire: A Day on Earth in Nature and Imagination.* New York: Viking, 2007.

CHAPTER 4

Ellis, Lee, and Eshah A. Wahab. "Religiosity and Fear of Death: A Theory-Oriented Review of the Empirical Literature." *Review of Religious Research* 55, no. 1 (March 2013): 149-189.

Francis (Pope). *Encyclical on Climate Change and Inequality: On Care for Our Common Home.* New York: Melville House Books, 2013.

Jung, Carl. "The Soul and Death." In *The Collected Works of C.G. Jung.* 2nd ed.. Princeton, NJ: Princeton University Press, [1934] 1969.

McPhee, John. *The Control of Nature.* New York: Farrar, Straus and Giroux, 1989.

Nash, Roderick. *Wilderness and the American Mind.* New Haven, CT: Yale University Press, 1965.

"The Quest for Immortality." *New Scientist* 216 (October 20, 2012): 38-40.

CHAPTER 5

Bain, Erin M., Lucas Habib, and Stan Boutin. "Impacts of Chronic Anthropogenic Noise from Energy-Sector Activity on Abundance of Songbirds in the Boreal Forest." *Conservation Biology* 22, no. 5 (July 2008): 1186-1193.

Balcomb, Kenneth C. "Letter to J.S. Johnson, SURTASS LFA Sonar OEIS/EIS Program Manager." February 23, 2001.

Baldwin, Ann Linda. "Effects of Noise on Rodent Physiology." *International Journal of Comparative Psychology* 20 (2007): 134-144.

Barber, Jesse R., Kevin R. Crooks, and Kurt M. Fristrup. "The Costs of Chronic Noise Exposure for Terrestrial Organisms." *Trends in Ecology and Evolution* 25, no. 3 (2009): 180-189.

Beatley, Timothy. *Handbook of Biophilic City Planning and Design.* Washington, DC: Island Press, 2016.

Clark, Charlotte, and Stephen A. Stansfeld. "The Effect of Transportation Noise on Health and Cognitive Development: A Review of Recent Evidence." *International Journal of Comparative Psychology* 20, no. 2 (2007): 145-158.

Coghlan, Andy. "Dying for Some Quiet: The Truth About Noise Pollution." *New Scientist,* August 22, 2007.

Francis, Clinton D., Catherine P. Ortega, and Alexander Cruz. "Noise Pollution Changes Avian Communities and Species Interactions." *Current Biology* 19 (2009): 1415-1419.

Fuller, Richard A., Phillip H. Warren, and Kevin J. Gaston. "Daytime Noise Predicts Nocturnal Singing in Urban Robins." *Biology Letters* 3 (August 2007): 368-370.

Greiser, Eberhard, Claudia Greiser, and Katrin Janhsen. "Night-time Aircraft Noise Increases Prevalence of Prescriptions of Antihypertensive and Cardiovascular Drugs Irrespective of Social Class: The Cologne-Bonn Airport Study." *Journal of Public Health* 15, no. 5 (September 2007): 327–337.

Haralabidis, Alexandros S., et. al. "Acute Effects of Night-time Noise Exposure on Blood Pressure in Populations Living Near Airports." *European Heart Journal,* 2008. doi:10.1093/eurheartj/ehn013.

Hempton, Gordon, and John Grossman. *One Square Inch of Silence: One Man's Quest to Preserve Quiet.* New York: Atria Books, 2010.

Krause, Bernie. *The Great Animal Orchestra: Finding the Origins of Music in the World's Wild Places.* New York: Back Bay Books, 2013.

Montgomerie, Robert, and Patrick J. Weatherhead. "How Robins Find Worms." *Animal Behavior* 54, no. 1 (July 1997): 143-151.

Moore, Kathleen Dean, and Michael Nelson. *Moral Ground: Ethical Action for a Planet in Peril.* San Antonio, TX: Trinity University Press, 2010.

Riitters, Kurt H., and James D. Wickham. "How Far to the Nearest Road?" *Frontiers in Ecology* 1 (April 2003): 125-129.

Schafer, R. Murray. *Tuning of the World.* New York: Random House, 1977.

Schueck, Linda S., John M. Marzluff, and Karen Steenhof. "Influence of Military Activities on Raptor Abundance and Behavior." *The Condor* 103, no. 3 (August 2001): 606-615.

Slabbekoom, Hans, and Ardi den Boer-Visser. "Cities Change the Songs of Birds." *Current Biology* 16 (2006): 2326-2331.

Watson, Chris. *A Small Slice of Tranquility.* London: BBC Radio program, 2003.

Weilgart, L. S. "The Impacts of Anthropogenic Ocean Noise on Cetaceans and Implications for Management." *Canadian Journal of Zoology* 85 (2007): 1091-1116.

Wood, William E., and Stephen M. Yezerinac. "Song Sparrow (Melospiza Melodia) Song Varies with Urban Noise." *The Auk* 123 (2006): 650-659.

Wright, Andrew J., et. al. "Do Marine Mammals Experience Stress Related to Anthropogenic Noise?" *International Journal of Comparative Psychology* 20, no. 2 (2007): 274-231.

CHAPTER 6

"American Lawn Is Now the Largest Single 'Crop' in the U.S." *WorldPost,* August 17, 2015.

Bormann, Herbert F., Diana Balmori, and Gordon T. Beballe. *Redesigning the American Lawn: A Search for Environmental Harmony.* New Haven, CT: Yale University Press, 2001.

Corbett, Julia B. "Blooming." In *Seven Summers: A Naturalist Homesteads in the Modern West.* Salt Lake City: University of Utah Press, 2013.

Haeg, Fritz. *Edible Estates: Attack on the Front Lawn.* New York: Metropolis Books, 2008.

Pollan, Michael. *Second Nature: A Gardener's Education.* New York: Grove Press, 2003.

Robbins, Paul. *Lawn People: How Grasses, Weeds, and Chemicals Make Us Who We Are.* Philadelphia: Temple University Press, 2007.

References

CHAPTER 7

Allenby, Brad. "Earth Systems Engineering and Management: A Manifesto." *Environmental Science and Technology,* December 1, 2007, 7960-7965.

Berry, Wendell. *Life Is a Miracle.* Berkeley, CA: Counterpoint Press, 2000.

Bertolas, Randy. "Cross-Cultural Environmental Perception of Wilderness." *Professional Geographer* 50 (February 1998): 98-111.

Boroditsky, Lera. "Linguistic Relativity." In *Encyclopedia of Cognitive Science.* Hoboken, NJ: Wiley and Sons, 2005. https://login.ezproxy.lib.utah.edu/login?url=https://search .credoreference.com/content/entry/wileycs/linguistic_relativity/0?institutionId=6487.

Evernden, Neil. "Beyond Ecology: Self, Place, and the Pathetic Fallacy." *North American Review* 263, no. 1 (1978): 16-20.

Foster, Charles. *Being a Beast: Adventures Across the Species Divide.* New York: Metropolitan Books, 2016.

Hanh, Thich Nhat. *Love Letter to the Earth.* Berkeley, CA: Parallax Press, 2013.

Kimmerer, Robin Wall. *Braiding Sweetgrass: Indigenous Wisdom, Scientific Knowledge, and the Teachings of Plants.* Minneapolis: Milkweed Editions, 2013.

———. "Speaking of Nature." *Orion* (March/April 2017): 14-27.

Larson, Brendon. *Metaphors for Environmental Sustainability.* New Haven, CT: Yale University Press, 2011.

Leopold, Aldo. *Sand County Almanac.* New York: Oxford University Press, 1949.

McPhee, John. *The Control of Nature.* New York: Farrar, Straus and Giroux, 1989.

Marston, Betsy. "Nature's Disappearing Vocabulary." In "Heard Around the West" column in *High Country News,* January 23, 2017. https://www.hcn.org/issues/49.1/ natures-disappearing-vocabulary-bobcat-decoys-setting-the-barr-for-climate-data.

Muir, John. *My First Summer in the Sierra.* Boston and New York: Houghton-Mifflin, 1911.

Orr, David W. *Ecological Literacy: Education and the Transition to a Postmodern World.* Albany: State University of New York Press, 1992.

Orrell, David. *The Future of Everything: The Science of Prediction.* New York: Basic Books, 2008.

Robock, Alan. "20 Reason Why Geoengineering May Be a Bad Idea." *Bulletin of the Atomic Scientists* (May/June 2008): 14-18, 59.

Roeser, Sabine. "Risk Communication, Public Engagement, and Climate Change: A Role for Emotions." *Risk Analysis* 32, no. 6 (2012): 1033–1040.

Rolston, Holmes III. "Does Nature Have Intrinsic Value?" In *Environmental Ethics: Readings in Theory and Application,* edited by Louis P. Pojman and Paul Pojman, 107–120. Belmont, CA: Wadsworth, 2008.

Sanders, Scott Russell. "Kinship and Kindness: On Deepening Our Connection with our Fellow Beings." *Orion* (May/June 2016): 26-35.

Tostevin, Bob. *The Promethean Illusion: The Western Belief in Human Mastery of Nature.* Jefferson, NC: McFarland Books, 2011.

Whyte, K. P, Brewer, J. P., and Johnson, J. T. "Weaving Indigenous Science, Protocols, and Sustainability Science." *Sustainability Science* 11 (April 2015): 25-32.

References

CHAPTER 8

Alexander, Thomas G. "Irrigating the Mormon Heartland: The Operation of the Irrigation Companies in Wasatch Oasis Communities, 1847-1880." *Agricultural History* 76 (Spring 2002): 172-187.

Benjamin, Walter. "Paris, Capital of the Nineteenth Century." In *Reflections: Essays, Aphorisms, Autobiographical Writings*, 146. New York: Schocken, 1986.

———. *The Arcades Project.* Cambridge, MA: Belknap, 1999.

Bradley, Martha S. *ZCMI: America's First Department Store.* Salt Lake City: ZCMI, 1991.

Foltz, Richard C. "Mormon Values and the Utah Environment." *Worldviews* 4 (2000): 1-19.

Hine, Thomas. *I Want That! How We All Became Shoppers.* New York: HarperCollins, 2002.

Hooton, LeRoy W., Jr. "City Creek: Salt Lake City's First Water Supply." *Journal of History, LDS Archives* (n.d.).

Ritzer, George. *Enchanting a Disenchanted World: Continuity and Change in the Cathedrals of Consumption.* Thousand Oaks, CA: Sage, 1999.

Watson, Thora. *The Stream That Built a City: History of City Creek, Memory Grove, and City Creek Canyon Park, Salt Lake City, Utah.* Salt Lake City: T. Watson, 1995.

Worster, Donald E. *Rivers of Empire: Water, Aridity, and the Growth of the American West.* New York: Pantheon Books, 1986.

CHAPTER 9

Cornelissen, Tatiana. "Climate Change and Its Effects on Terrestrial Insects and Herbivory Patterns." *Neotropical Entomology* 40, no. 2 (April 2011): 155-163.

Eisner, Thomas. *For Love of Insects.* Cambridge, MA: Harvard University Press, 2003.

Glausiusz, Josie. *Buzz: The Intimate Bond Between Humans and Insects.* San Francisco: Chronicle Books, 2004.

Granado, Laura Carmilo, Ronald Ranvaud, and Javier Ropero Peláez. "A Spiderless Arachnophobia Therapy: Comparison Between Placebo and Treatment Groups and Six-Month Follow-up Study." *Neural Plasticity* (July 2007): 1-11. doi:10.1155/2007/10241.

Hillman, J. 1991. *Going Bugs.* Spring Audio, Gracie Station, New York.

Hunter, Mary Carol R., and Mark D. Hunter. "Designing for Conversation of Insects in the Built Environment." *Insect Conservation and Diversity* 1 (October 2008): 189-196.

Kellert, Stephen R. "Affective, Cognitive, and Evaluation Perceptions of Animals." In *Behavior and the Natural Environment*, edited by Irwin Altman and Joachim F. Wohlwill, 117-151. New York: Plenum Press, 1983.

———. "Values and Perceptions of Invertebrates." *Conservation Biology* 7 (December 1993): 845-855.

Morris, Brian. *Insects and Human Life.* New York: Berg, 2004.

Pyle, Robert Michael. *Letting the Flies Out.* Gray's River, WA: New Riverside Press, 2013.

Russell, Sharman Apt. *Diary of a Citizen Scientist: Chasing Tiger Beetles and Other New Ways of Engaging the World.* Corvallis: Oregon State University Press, 2014.

Samways, Michael. *Insect Diversity Conservation.* Cambridge: Cambridge University Press, 2005.

References

Waldbauer, Gilbert. *How Not to Be Eaten: The Insects Fight Back.* Berkeley: University of California Press, 2013.

Wilson, E. 0. "The Little Things That Run the World," *Conservation Biology* 1, no. 4 (December 1987): 344-346.

CHAPTER 10

Annual Energy Review 2010. Washington, DC: U.S. Energy Information Administration, October 2011. www.eia.gov/aer.

Brandon, Gwendolyn, and Alan Lewis. "Reducing Household Energy Consumption." *Journal of Environmental Psychology* 19, no. 1 (March 1999): 75-85.

Chambless, Ross S. *Plugging into Nature: A Radio Journey to Find the Sources of My Electricity.* MA thesis, University of Utah, May 2011.

Close the Door. Last accessed February 10, 2018. *www.closethedoor.org.uk/.*

Guerin, Denise A., Becky L. Yust, and Julie G. Coopet. "Occupant Predictors of Household Energy Behavior and Consumption Change as Found in Energy Studies Since 1975." *Family and Consumer Science Research Journal* 29 (November 2000): 48-80.

Hill, Holly. *Food Miles: Background and Marketing. National Sustainable Agriculture Information Service,* 2008.

Holden, Erling, and Kristin Linnerud. "Environmental Attitudes and Household Consumption: An Ambiguous Relationship." *International Journal of Sustainable Development* 13, no. 3 (2010): 217-231.

Jensen, Derrick. "Forget Shorter Showers: Why Personal Change Does Not Equal Political Change." *Orion* (July/August 2009): 18-19.

Lord, Barry. *Art and Energy: How Culture Changes.* Washington, DC: AAM Press, 2014.

Machiavelli, Niccolò. *The Prince.* New York: Oxford University Press, [1532] 1952.

Miranowski, John A. "Energy Consumption in US Agriculture." In *Agriculture as a Producer and Consumer of Energy,* 68-111. Cambridge, MA: CABI Publishing. 2006.

Schultz, P. Wesley, Jessica M. Nolan, Robert B. Cialdini, Noah J. Goldstein, and Vladas Griskevicius. "The Constructive, Destructive, and Reconstructive Power of Social Norms." *Psychological Science* 18 (May 2007): 105-109.

"The Tricky Truth About Food Miles." *Shrink That Footprint. www.shrinkthatfootprint.com.* Accessed June 8, 2017.

Wallenborn, Grégoire, Marco Orsini, and Jeremie Venhaverbeke. "Household Appropriation of Electricity Monitors." *International Journal of Consumer Studies* 35 (February 2011): 146-152.

EPILOGUE

Freyfogle, Eric T. *A Good That Transcends: How U.S. Culture Undermines Environmental Reform.* Chicago: University of Chicago Press, 2017.

Oliver, Mary. "Mindful." In *Why I Wake Early: New Poems,* 58. Boston: Beacon Press, 2004.

Watts, Alan. Twitter message. @AlanWattsDaily. July 4, 2013.

About the Author

JULIA CORBETT is a professor in the Department of Communication and the Environmental Humanities Graduate Program at the University of Utah, Salt Lake City. She authored one of the first texts in environmental communication, *Communicating Nature: How We Create and Understand Environmental Messages.* Her second book, *Seven Summers: A Naturalist Homesteads in the Modern West,* "enacts the insights of feminist nature criticism" in a memoir about a small cabin in a wild place in a twenty-first-century landscape under acute pressure; it was short-listed for the Reading the West Award. Her environmental nonfiction essays have been published in venues such as *Orion, High Country News,* and *Camas.* Her current book project and much of her teaching involve communicating climate change. Before receiving her MA and PhD at the University of Minnesota in 1994, she was a newspaper reporter, a park ranger, a naturalist, a natural resources information officer, and a deputy press secretary. She summers in the mountains of western Wyoming in her cabin.